New Tech, New Ties

New Tech, New Ties

How Mobile Communication Is Reshaping Social Cohesion

Rich Ling

The MIT Press
Cambridge, Massachusetts
London, England

For information on quantity discounts, email special_sales@mitpress.mit.edu.

Set in Sabon by SPi, Pondicherry, India. Printed and bound in the United States of America.

Library of Congress Cataloging-in-Publication Data

Ling, Richard Seyler.
New tech, new ties : how mobile communication is reshaping social cohesion / Rich Ling.
p. cm.
Includes bibliographical references and index.
ISBN 978-0-262-12297-9 (hardcover : alk. paper)
1. Cellular telephones—Social aspects. 2. Interpersonal communication—Technological innovations—Social aspects. 3. Communication and culture.
I. Title.
HE9713.M43 2008
303.48'33—dc22
2007022315

10 9 8 7 6 5 4 3 2 1

to Marit, Nora, and Emma, and to my Aunt Anne for her kindness
and her resolve

Contents

Preface

Where I work, we have a tradition, or perhaps a ritual. Its functioning illustrates some of the main themes of this book: the relationship between ritual and social cohesion, with some aspects of mediated interaction in the mix.

Every Friday at 2 P.M., we gather and have a lottery for four or five bottles of wine. There are two rules: a person cannot win more than one bottle, and the winner of the last bottle is responsible for organizing the event the following week. We have several books of lottery tickets in various colors, and for 10 kroner each (equal to about 60 cents) one may buy as many chances as one wishes.

During the time that we are purchasing tickets, there is a lot of discussion as to which color is the lucky one. Some people buy only one ticket; some buy five or six. Some people buy only yellow tickets; some buy one of each color. The color selection and the number purchased by different people are topics of common discussion. After the tickets are bought, we sit in anticipation of the drawing. Some people make side deals by swapping tickets and discussing schemes for increasing their chances. Some people who do not have any small change on them borrow from the common bank for the lottery, and later there are loud and good-natured efforts at calling in the loans with threats of usurious interest rates.

People who are away on Friday afternoon often send text messages or emails to colleagues with requests to spot them the money for some lottery tickets, and colleagues who have been as far afield as Pakistan and the United States have won in the lottery. Upon winning, these remote participants receive a text message that advises them of their good luck. In addition, the messages often contain threats that the prize will be consumed before their return.

We always draw first for the least expensive bottle and last for the most expensive one. If someone has won a less expensive bottle but then has the fortune of winning a more expensive one, he or she may exchange the cheaper bottle for the more expensive one, and we then have a new drawing for the bottle that has been put back on the table.

In the worst case, the rules can result in someone's winning an inferior bottle of wine but still having to organize the following week's lottery. This happens when someone who has won a less expensive bottle also wins the most expensive one and puts the cheaper bottle back on the table for a new drawing. Since this cheaper bottle is the last bottle on the table, the person winning that one also has the responsibility of organizing the lottery the following week. Thus, winning this bottle is tinged with unwanted acquisition of responsibility. If the winner does not follow through on the responsibility, it is a serious breach of norms.

The wine lottery is anticipated. It marks the ending of the work week. It gives us an informal opportunity to gather and chat and to have one last laugh together before the weekend. It generates a certain group history, and the various intrigues and unanticipated outcomes are a small but important element in the flux of our interaction. In sum, the event helps to tie us together as a group.

We collectively engineer a lighthearted mood, and in that engineering of a mood we connect in a way that deepens our sense of group identity. The staging of the event, the joy or faux disappointment at the outcome of the drawing, and the petty but transparent attempts to jigger the results are all parts of the play. There are jeers when only the women or only the people on one side of the table win. There are assertions that absent participants have won empty bottles, and there is false anguish when the winning ticket is only one digit away from the ticket someone is holding. These elements are also how we engage in the generation of good will and group identity. We remember that K___ wins too often for our taste, or that T___ is all too conservative in her purchasing of tickets. Our mutual recognition of a common mood is the core of the tradition, and it helps to generate a common sense of identity and cohesion.

Our weekly wine lottery is a ritual. The event is a ritual in the more common sense of the word, since it happens with certain regularity, it follows a certain form, and the process of staging it is similar from week to week. The more interesting part, however, is the Durkheimian/

Goffmanian sense of ritual that is played out here. Rather than looking at the regularity of the event or the liturgy used in the staging of the affair, it is important to look at the sense of mutual recognized engagement that arises when we gather. This, in turn, generates a common sense of what Durkheim calls "effervescence." It is through this kind of process that the group develops its sense of identity. The more serious tones of the work week are cast aside and levity takes center stage. I know that my colleagues are there for a laugh, as am I. It gives us something to talk about, and it helps to define who is a member of the group and who is not.

Like many other rituals, the wine lottery is based largely on co-present interaction. But the line between the co-present and the non-present is becoming less distinct. On occasion, colleagues who are in a meeting in another portion of the building or those who are in Kuala Lumpur, Dhaka, or New York have participated, albeit somewhat indirectly.

Thus, in this seemingly banal event, we have, in a neat little package, ritual interaction, cohesion, and mediated communication.

Mobile communication allows us to participate in social interactions that were previously reserved for only those who were physically present. This is not to say that the wine lottery is about to become entirely virtual. The real core of the tradition is the co-present interaction. Without that interaction, there would be no atmosphere, no excitement, no effervescence. But participation has been given a new dimension. The absent colleague is a part of the interaction. The interactions the absent colleague must go through to secure a loan from a present colleague and the good-natured insults and threats the absent colleague must endure should he or she win have become parts of the interaction.

It is enticing to say that our wine lottery exemplifies certain changes that are happening in society. Mobile communication is being used in the pursuit of romance, in the coordination of families, in the exchanging of humor and gossip, and in many other daily situations. In each case, there are ritual forms, there is reliance on co-present understandings, and there is development—and sometimes erosion—of social cohesion. These are the themes I examine in this book.

Acknowledgements

First, I would like to recognize those people at Telenor Research and Development (now Telenor Research and Innovation) who have provide me the institutional space to think through the issues presented here. First on the list is Birgitte Yttri, my former writing partner and now my boss. I am sorry to have lost her as a direct colleague but grateful for her continued support of my activities. She, Marianne Jensen, Kristin Braa, Kristin Thrane, and our institute leader Hans Christian Haugli have been generous in supporting my work on this book.

I would also like to give special thanks to Susan Douglas and Mike Traugott at my other institutional home, the Department of Communication Studies at the University of Michigan. Their support and their willingness to take me on as a somewhat virtual colleague are much appreciated. In addition I thank Constance C. and Arnold H. Pohs for their support.

I want to thank others those who have been quite directly involved in the development and writing of this book. Early ideas associated with the book arose in papers that I have written for conferences organized by Shin Dong Kim and James Katz. These conferences gave me the motivation and the opportunity to start thinking about the ideas of ritual, mobile communication, and social cohesion. I also want to thank them on a personal level for their friendship and collegiality.

Other early influences on the book arose in the E-living project chaired by Ben Anderson and the SOCQUIT project chaired by Jeroen Heres. In each case, we were able to examine the issues of social capital and its interaction with mobile communication. Again, as with the conference papers, these projects gave me the opportunity to work out ideas that later were drawn into the pages of this book. Thanks to Ben, Jeroen, and all the other participants in these projects.

Another person who deserves recognition is Russ Neumann at the University of Michigan. He welcomed me into his graduate seminar as a slightly irreverent colleague, and he provided me with good comments on the early work associated with the book.

Several people have helped me in quite direct ways. First, I would like to thank my Telenor colleagues Tom Julsrud, John Willy Bakke, Hege Andersen, and Linn Prøitz. Tom has kept me abreast of ideas such as that of tacit knowledge in groups. John Willy is always available to help me think through different intellectual snarls. In the case of this book, the section on the difference between ritual as an obsessive behavior and ritual as the motor of social cohesion came from a chat we had a lunch several months ago. Hege's and Linn's work on various aspects of flirting, romance, and sexuality carried out via text messaging was invaluable in the development of chapter 8.

Scott Campbell, Ron Rice, Naomi Baron, and Leslie Haddon read and commented on chapters or helped me work out ideas. Scott read and commented on the entire manuscript, gave me valuable ideas, and saved me from guaranteed embarrassment by pointing out literary, factual, and grammatical indiscretions. Ron deserves a Nobel Prize for his careful review of chapter 8. My interaction with Naomi on the role of texting has been a wonderful and intellectually exciting adventure. Leslie has been a great inspiration. He has a broad and well-founded sociological insight and he is always able to help me think about issues and questions that need addressing. I would also like to thank Leopolda Fortunati, Christian Licoppe, and Mizuko Ito for their collegiality and their important insights.

Special thanks to Doug Sery of The MIT Press for his guidance, assistance, and trust, and also to Paul Bethge.

I also want to make a special point of thanking Randall Collins. His work has been a central inspiration for the book, and he has done me the service of reviewing the manuscript. As I note in chapter 2, it was his insights into the more "interactionist" side of Durkheim that really led me to the notion of ritual and mobile communication. His lucid and astute understanding of both Durkheim and Goffman was central to the development of my thinking on these issues. His own work on interaction ritual chains brought much of this material together. I am honored that he was willing to help foster my ideas on mediated ritual interaction. He took the time to assist me in thinking about some of these issues, some of which are oblique to his own thinking. His willingness to engage in this discussion shows a greatness of spirit. I am grateful for his many useful comments regarding the actual manuscript. They have helped me to bring out various aspects of mediated ritual interaction.

Finally, I want to thank my wife Marit for her support while I have worked on this book and for her ability to ground my ideas in everyday reality. Not virtual reality, but real reality. I also need to thank my daughters Nora and Emma for keeping me young and in touch with the ideas and directions of their generation. Lizzom takk og krämmar.

1
Mobile Communication and Ritual Interaction: The Plumber's Entrance

Sometimes, remarkable things happen right in front of your nose. One morning at about 8:30, I was bidding farewell to some guests who had spent the night at our home in Oslo. As it was a pleasant day, I stood on our porch with the front door open as we said our final goodbyes. At this point, a plumber, with whom we had an appointment, appeared around the corner of the house. He was checking the address against an order he held in one hand. In his other hand was a mobile phone, into which he was talking. Indeed he seemed quite engaged in the phone conversation.

The plumber, who I had not met before, continued his phone conversation, checked the address against that which was written on his order, and walked past the departing guests and up the two steps to our porch. Without breaking stride, he walked past me and into the house, giving me a minimal nod. He stopped in the vestibule and took off his shoes—as is common in Norway—and then continued down the hallway into the kitchen. All the while, he continued his phone conversation. I exchanged a bemused glance with my departing guests. After a few closing comments, I turned and retreated into the kitchen, where I was received by the plumber, who by this time had completed his telephone conversation.

Aside from ruffled feathers, I soon nearly forgot the episode. After all, it was just one more example of how the mobile telephone has entered into our daily lives. People talk on them in the most unexpected places and talk about the most unexpected issues. Eventually, however, the situation struck me as more interesting. There was a layering of co-present and mediated interactions being worked out, to the advantage of some and the surprise of others.

The plumber's entrance disturbed the farewells I was exchanging with my friends. With a little quick thinking and some social polish, he could have easily avoided the awkward situation. Had he interrupted his telephonic interaction for a moment and quickly introduced himself, the situation would have been far less remarkable. Had he paused outside the door while he finished the telephonic interaction and I finished my parting, it would not have been remembered. Had he read the situation (he was, for the first time, entering into the private sphere of another person) and provided the proper obeisance, the event would have been of little note, certainly not worthy of an introduction to a book.

As it was, however, the conversation between the plumber and his interlocutor was not interrupted. That event continued unabated. The situation did, however, make a travesty of the traditional greeting we give when guests enter our home. There was not the exchange of greetings, handshakes, small talk, and the extension of an invitation into the home. Indeed, about the only residues of that were the slight nod and the removal of the shoes. All the other elements of the standard exchange were neatly trumped by the telephonic interaction.

From the plumber's perspective, the phone call was obviously an important event in his life. I did not overhear the topic, but it might have been his boss giving him orders for the day; it might have been the personnel at his child's day-care center telling him that his child was sick; it might have been a buddy with whom he was planning weekend activities; it might have been his wife telling him that she wanted to leave him. I was left to speculate about this. The mediated form of interaction, however, was engrossing to the point that he judged that he could place the perhaps more routine co-present interaction into background, particularly insofar as I was ostensibly more concerned with dispatching my guests. Because of this, the plumber's visit got off on the wrong foot. To be sure, the best thing that came from the encounter was the repair of a leaky faucet and a good anecdote with which to hook the interest of the reader. Unfortunately, I had to pay dearly for this "hook" with a rather weighty bill from the plumbing company.

■

This book is about how the mobile telephone affects our sense of social solidarity. Does it contribute to our sense of social cohesion, or does it detract from it?[1]

As in the example of the plumber's greeting, I look at how the mobile telephone plays into both co-present and mediated interaction. I find that mobile communication seems to strengthen communication within the circle of friends and the family. The material presented here indicates that it supports better contact within the personal sphere, sometimes at the expense of interaction with those who are co-present. My assertion is that it allows for tying of tighter bonds via various forms of ritual interaction.

There are many consequences of mobile communication. These include extending our sense of safety and security, the ability to micro-coordinate, the disturbance of the public sphere, and changing the way teens and parents experience the emancipation process (Ling 2004b).

Mobile communication has altered the way that social situations develop and the way that they are carried off. Before the development of telecommunications, we could converse with only those who were near at hand. It was possible to correspond, of course, with letters, but these were widely asynchronous. With the development of the telegraph, message transmission was nearly simultaneous, at least between telegraph offices. The final delivery of a message relied on the transport of a piece of paper. The telephone extended this kind of simultaneity into the home and to specific geographic locations (offices, phone booths, etc.). Mobile communication, however, has meant that we can talk to others regardless of where they are.

Mobile communication is different from other forms of interpersonal mediation (PC-based email and instant messaging, for example[2]) in that mobile telephony makes us each personally addressable. We are perpetually accessible to friends, family members, acquaintances, and even people with whom we might not want to speak at the moment. With mobile communication, we call to individuals, not to locations.

With traditional land-line telephony, we called to specific places in the sometimes misbegotten hope that our intended interlocutor would be somewhere near the phone we were calling—perhaps upstairs, out in the yard, or in another room on the same floor. We might chat with the person who answered the phone, who then would look for the requested person. After a short interlude, he or she would "come to the phone" and we could get down to the issue at hand. Such is not the case with mobile telephony. I call to the individual. Where he or she is and what he or she is doing may be a surprise to both parties. Bathrooms, funeral services, sporting events, art galleries, refined restaurants, and sexual trysts are all "in play" in what the sociologist in me wants to call "telephonic interaction contexts." All of this affects how we develop, maintain, and dispose of social cohesion. It affects how often we interact with others, what we say, and what we know about one another. Licoppe (2004) calls this "connected presence." Katz and Aakhus (2002b) see it as a dimension of "apparatgeist." Ito and Okabe (2005) write:

> . . . mobile phones do undermine prior definitions of social situations, but they also define new technosocial situations and new boundaries of identity and place. To say that mobile phones cross boundaries, heighten accessibility, and fragment social life is to see only one side of the dynamic social reconfigurations heralded by mobile communications. Mobile phones create new kinds of bounded places that merge infrastructures of geography and technology, as well as technosocial practices that merge technical standards and social norms.

More than other forms of mediated interaction, mobile communication favors contact with those with whom we are close—family members, friends, colleagues. Since we are always accessible, we have the ability to play on and develop those relationships, perhaps at the expense of what Granovetter (1973) calls "loose ties."

The mobile telephone has provided us with an opportunity to examine the reaction of society to the adoption and use of a new communication technology. We have the opportunity to examine how this technology has influenced sociation. We have been witnessing a transition from geographically bound telecommunication to personal addressability via mobile communication. This transition provides us with the chance to examine how the introduction of a technology affects the development of, or the destruction of, social cohesion (Ling et al. 2002, 2003; Ling 2004a–d).

■

There is the general sense that in contemporary society we have fewer intimate ties. In the United States, according to McPherson et al. (2006), social networks were smaller in 2004 than in 1985, with nearly three times as many people saying there was no one with whom they discussed important matters. Putnam (2000), also considering the United States, examined the loss of social capital relative to the period immediately after World War II.[3]

During the time frames that McPherson et al. and Putnam examined, many more people began using mediated communication (email, list serves, networking sites such as MySpace and Facebook, chat rooms, text messages, mobile telephony, instant messaging) to contact close members of their networks (Boase et al. 2006). There is evidence that, in the words of Yeats, "the center will not hold." Many have suggested that interactive technology is straining the social structure and causing a drift toward individualism. Kraut (1998) and Nie (2001) think there is greater individualism; Katz and Rice (2002) claim the opposite.

The situation with the plumber exemplifies how mediated interaction can sometimes take precedence over the co-present. It shows how face-to-face rituals can be set into the background when the flux of a mobile phone conversation might demand that. On a more positive note, if we give the plumber the benefit of the doubt, the example might show how the device works against what others have seen as the increasing isolation and insularity of present-day society. He was after all, able to have an engrossing conversation with a friend or a colleague. This might be a sign that the technology is providing the means for more, not less, social cohesion. Indeed, Katz et al. suggest this in their analysis of mediated interaction and the concept of community (2004a).

The rise of mobile communication has given us cause to think about the mechanics of social interaction. What, for example, are the means by which the social ties are being tightened or loosened? How do we develop, maintain, and manage the social order in our everyday lives? What are the social consequences of mobile communication for interaction with friends and family? Is this interaction carried over into the ideational world? The assertion here is that via the use of ritual—and ritual in the very specific sense developed by Durkheim and elaborated by Goffman and Collins—we develop social cohesion.[4]

•

In an era of broad discussion about the effects of communication technologies on social cohesion, ritual and its use in the context of mediated interaction is an important topic. In this book, I develop the idea that it is ritual interaction—either mediated or co-present—that leads to social cohesion. It is the glue that holds society together.[5]

The sociologist Emile Durkheim provided the central insight into how ritual provides the catharsis underlying social cohesion. Durkheim's analysis of Aboriginal rites (1995) gave him the insight that the main product of this situation is the rejuvenation of social solidarity. Erving Goffman (1967) and Randall Collins (2004a) took Durkheim's insight and applied it to everyday situations.

Neither Durkheim nor Goffman nor Collins examined mediated interaction. Durkheim worked in a period before the broad adoption of telephony and described a society for which access to this technology was far in the future. Goffman was somewhat open to the examination of mediated interaction, but his main focus was clearly on the co-present. Collins is explicit in saying that to engender cohesion a ritual cannot be mediated. Indeed, he says, physical co-presence is a requisite aspect of ritual interaction.

How does ritual transcend the co-present, and indeed the co-temporal? To what degree does ritual interaction transcend the here and now? Are we able to use the devices of mobile communication to bring about social solidarity?

I am not suggesting that co-present interaction is losing its place as a central context for the development of social cohesion. I am not suggesting the rise of a cyber-utopia in which the most important interaction is carried out in bits and bytes. We are not on the cusp of era ruled by simple mediated interaction. It is clear that face-to-face or body-to-body (Fortunati 2005) interaction is the most efficient way of generating Collins's focused emotions and mutual recognition. In spite of this, as evidenced by my interaction with the plumber, mediated interaction is taking a larger slice of the pie. We are using this form of interaction more often and for more things. We send emails, SMS messages,[6] and even multimedia messages. Through these forms of mediation, we share photos and jokes and we flirt. Our "face time" is being extended and embroidered by mediated interaction. We are anticipating our get-togethers with one another and following up our meetings with various messages and remembrances. They help us to maintain our social contacts, and they modify, readjust, and displace social interaction. Although the co-present is still the locus of social ritual, it is being extended and modified by mediated interaction.

In contrast to the assertions of Collins, I will explore how mediated contact via mobile telephone can also be seen in the context of ritual interaction. The way we greet one another over the phone, the way we relate stories, and the way we use the phone to organize our daily life show that, in many respects, ritual interaction can be carried out via interactive media. Further, particular forms of interaction and parlance that seem to occur only via mobile phone can be seen as mediated ritual interaction. Thus, I am interested in expanding Collins's sense of ritual interaction into the area of mediated communication.

The Many Meanings of the Word 'Ritual'

A central idea of this book is that ritual has an integrative effect. A look at the various meanings of the word 'ritual' indicates that it has a varied past. Indeed the word is burdened with several sometimes conflated meanings. On the whole, many see the word as more pejorative than positive, in that ritual can be seen as a set of actions performed more for their symbolic or prescribed value than for any real effect. Since 'ritual' is used in a mostly positive sense here, it is important to parse the meaning of the word. There are psychological senses and group senses. The sense taken in this book is that of a catalyst for cohesion.

In the psychological sense, ritual is sometimes seen as a repetitive or, in the worst case, a compulsive habit. It can be a completely inflexible behavior that is seen as necessary in the ordering of one's day. A loose sense of this definition of 'ritual' is a series of actions that help us to move through our lives. We may have personal rituals—for example, taking a walk on Saturday; showering in the morning; reading the paper with a pot of coffee on Sunday; always locking the door, checking that the oven is turned off, and petting the cat before going to bed. At its extreme, this definition of 'ritual' can include obsessive behavior. A daily or perhaps hourly check to make sure that all our *National Geographic* magazines are in their proper order, with the maps in place and the bindings two inches from the edge of the shelf, might strike others as taxing if not disturbed.[7] This definition carries with it a sense of being inflexible, obsessive, and repetitive. Dulaney and Fiske (1994, p. 245) write:

Rituals often involve washing and other forms of purification, orientation to thresholds and boundaries and colors that have special significance. Rituals tend to involve precise spatial arrays and symmetrical patterns, stereotyped actions, repetitive sequences, rigidly scrupulous adherence to rules (and often the creation of new rules), and imperative measures to prevent harm and protect against immanent dangers. These features typify rituals but they also define a psychiatric illness, obsessive-compulsive disorder.[8]

Indeed, in discussions of obsessive-compulsive disorder, the behavior of patients is described as 'ritual'. Freud (1963) asserts that ritual is similar to obsessive compulsive behavior. Bell (1997, p. 13) comments:

In a 1907 essay . . . Freud drew a provocative comparison between obsessive activities of neurotics and those "religious observances by means of which the faithful give expression to their piety," such as prayers and invocations. For Freud the neurotic's innumerable round of little ceremonies, all of which must be done just so, as well as the anxiety and guilt that accompanies these acts, imply a similarity between the causes and religion and the causes of obsessional neuroses.

In summary, Freud suggested that ritual and religion are both rooted in repression and displacement. It is important to note that ritual in this sense will not be considered here.

■

Freud's linking of ritual in the psychological sense and the practice of religion provides a convenient segue to the next meaning of 'ritual', namely the execution of a group ritual in a closed and unthinking manner. This is when a group perhaps mindlessly follows a specific liturgy or process in which peer pressure seems to be more important than the essential meaning of the actions. The group ritual seems to have no clear connection to the alleged sense that the process will result in any positive outcome, such as redemption, beatification, bliss, or greater insight. Indeed, the first recorded use of the word 'ritual' carries this sense. According to the Oxford English Dictionary, this dates back to 1570 and describes ritual as "contayning no maner of doctrine . . . but onely certayn ritual decrees to no purpose." According to this sense, ritual can seem to be more facade than content. Ritual in this sense is a specific liturgy that includes the same series of actions performed in the same sequence. In this case, there may be a symbolic content of the actions that is often proscribed by a religion (Merton 1968). The nonbeliever uses the word 'ritual' to describe a congregation's genuflection, liturgy, and praying in unison. We also speak of ritual as a way of cleansing the soul or gaining absolution. There is the ritual bath as seen in the Christian baptism, the Hindu tirthayatra, the Muslim cleansing before prayer, the Hebrew mikvah, and the Shinto ablution.

On a less harmonious note, there is ritual murder or mutilation. In this case, as with the murders carried out by the Manson cult or the Incan sacrifice of a teen girl on the Nevado Ampato volcano, the ritual was, in the estimation of the cult members, supposed to initiate a social transition or to appease the gods. Where the former definition was perhaps more focused on the actions of the individual, this definition of 'ritual' is more social. In both cases, however, there is the obsessive ordering of events.

The obsessive notion of ritual—for perhaps obvious reasons—is often used despairingly. There can be the sense that the procedure is more form than function. Using the more "garden variety" meaning of 'ritual', and leaving the notion of ritual murder or mutilation aside, there is the sense that those who are involved in a group that bases a large part of its activities on specific repeated behaviors—such as dressing up in specific costumes or regalia and repeating the same order of service—are somehow giving a portion of their individuality over to the group. Indeed, those who are outside the circle may see this as a kind of false consciousness. It is obvious that there are elements of the obsessive sense of ritual contained in the notion of group ritual. There is the idea, at least on the part of the external observer, that the activity has no functional basis. There is the sense that the form of the process is more important than any broader outcome.

The confusion of this kind of ritual with obsessive behavior is also telling. The sense in both cases is that in some pathological way the individual derives ontological security from the exercise of the ritual. There is the sense that those who must become involved in these types of activities, be it a private ritual of checking on their *National Geographic* magazines or joining a group (such as the Grand Order of the Sacred Green Water Buffalos) that prescribes their behavior, are not well-functioning individuals. Ritual is a kind of crutch to help them through an otherwise confusing and difficult world. Though this may be an interesting path of research, it is not the one that will be taken here. Thus, I ask the reader to drop these two notions of ritual. This work is not about obsessive rituals, nor does it examine the sense of ritual seen in rote practices.

■

The sense of 'ritual' used here is that of a social phenomenon used in the process of sociation. It involves the establishment of a mutually recognized focus and mood among individuals, and it is a catalyst in the construction of social cohesion. The focus is not on obsessive or repetitive behavior, although ritual interaction can take place in these settings. Rather, the emphasis is on a group process and the outcome of that process.

In this case, ritual is not interesting in itself; rather, it is interesting because it provides insight into the functioning of society. The point is not to develop a taxonomy of rituals and to understand the differences between groups on the basis of that taxonomy. Rather, taking a page from Durkheim, the point is to examine how ritual is a social construction that has more or less ability to provide cohesion. Were it left at the group level, we would be missing an important sense of ritual interaction. It is in this area that we are indebted to Goffman and Collins, who extended the notion of ritual into the area of interpersonal interaction. "For Goffman," Collins writes (1998, p. 22), "every fleeting encounter is a little social order, a shared reality constructed by solidarity rituals which mark its entering and closing through formal gestures of greeting and departure, and by the little marks of respect which idealize selves and occasions."

According to Collins (ibid., p. 22), ritual is achieved when two or more people are physically assembled (a condition about which I will have much to say later) and when there is a mutually recognized focus on the same object or action—when there is a common mood or emotion that becomes intensified and there is, at some level, a shared sense of "effervescence" felt by the individuals.[9] The ritual is a catalyst that breaks down differences between the individuals and allows them to develop a sense of solidarity and allegiance to one another. The use of symbols, jargon, or totems allows the engagement generated by the event to be carried across time to other situations. This is because symbols, jargon, and totems are charged with intensity of the event and become reservoirs of shared energy that can be rejuvenated in further cycles of interaction. And this shared effervescence or engrossment marks the boundary between those who are inside the group and those who are outside.

It may be the case that ritual has no real functional purpose aside from making manifest and managing social cohesion. This statement is only partially true, however, since in many situations we draw ritual elements into the ongoing process of social interaction. Shaking hands, for example, has no real function. Shaking hands does not put meat on the table, nor does it do the work of daily life. However, unless there are handshakes at the outset of a meeting, the functional interaction gets off on the wrong foot. Refusal to shake hands can be seen as an affront and a snub. Ritual helps to channel interaction (Douglas and Isherwood 1979; Douglas 2003) and to develop a shared attitude that is integral in the management of conflict and the masking of inequality (Couldry 2003, p. 4; Bourdieu 1991).

In the following chapters, I will examine the scalability of ritual interaction when considering the differences between the larger Durkheimian and the more nuanced Goffmanian ritual (chapter 6). The Durkheimian notions of ritual can be used to examine the transition from one status to another (graduation, marriage, membership in a club or organization, etc.). In this sense, a ritual separates the transition process from mundane life. It marks the event as a special point worthy of attention. For example, Van Gennep

(2004/1960) notes that there is a liminality (or transition) associated with rites of passage: "[A]lthough a complete scheme of rites of passage theoretically includes preliminal rites (rites of separation) liminal rites (rites of transition) and post liminal rites (rites of incorporation) in specific instances these three types are not always equally important or equally elaborated."

According to Bourdieu's (1991) analysis of van Gennep, the significance of the transition is not in the shift from, for example, childhood into adulthood; rather, it is the dynamics of the event itself. (See also Couldry 2003.) In a ceremony of this kind, the individual experiences a social breach that is somehow resolved, and in the process he or she transcends from one level of social understanding to another. In this broader sort of ritual interaction (which, incidentally, is often choreographed by others), the individual crosses a threshold and is somehow transformed by the process.

In Goffmanian ritual, there is not the same palpable liminality when, for example, we shake hands with someone or when we invite a plumber into our home. Nonetheless, there is the sharing of a mood and a mutual recognition of the situation, albeit in miniature. These interactions are more difficult to observe, but we see them clearly when they fail.

Collins asserts that the interaction between ritual and social cohesion necessarily is co-present. He feels that there is the need to see the reaction of the others who are involved. There is the need to see how they give themselves over to the situation and let themselves be taken by its rhythms. There is often an element of power in these situations, since others (a priest, a rock star, a conversation leader) direct the flux and the pace of the situation. I do not quibble with the idea that co-present interaction is a profound basis of social solidarity. It has strongly liminal characteristics. I do, however, assert that social bonds can be maintained and nurtured through mediated interaction.[10] Once a bond is forged, however, mediated interaction is often as effective as co-present interaction in the development and maintenance of the interaction (Garton et al. 1997; Katz and Aspden 1997; Wellman et al. 2001).

■

So-called media rituals will not be examined here. These include irregular events such as the funeral of Princess Diana, regularly scheduled events such as the Olympic Games, the Super Bowl, and the finals of the UFEA Champions League, and daily events such as TV news programs (Selberg 1993; Couldry 2003). At the broadest level, these media events can be transformative. When we remember, for example, our personal situation when we learned of a dramatic event via the media, and thus there is the sense that we are participating in a broader social context. The classic recent example is, of course, the events of September 11, 2001.

There are obviously differences between the interpersonal (albeit mediated) ritual discussed here and the idea of media ritual as discussed by Couldry (2003). In this book I discuss the interpersonal, point-to-point form of interaction, whereas the term 'media ritual' often refers to a broadcast "one-to-many" form of interaction. The former is often motivated by personal need for interaction; the latter is often produced and sponsored by large commercial interests. This is not to say that people watching a football game or the marriage of a royal are not engaged in interpersonal interaction with the others who are co-locally watching with them. These broadcast events are also an element in developing the sense of team loyalty or nationhood (Couldry 2003; Selberg 1993; Selberg 1995). It is clear that the dynamics of ritual interaction are in play in these situations. However, they operate at another level of abstraction. They are not the same as the interpersonal interactions that arise in conversations, telephone calls, texting, instant messaging, and email.

Mobile Telephony in the World

In addition to ritual interaction and social cohesion, the other main topic of this book is mobile communication. The technical history of mobile communication devices dates back to the development of radio communication in the late 1800s. The mobile phone came into widespread use and acceptance in the mid 1990s. And in the period 1990–2005, the mobile phone achieved iconic status.

Mobile telephony is the child of radio communication and traditional land-line telephony. In 1910, in an early example of "mobile" telephony, Lars Magnus Ericsson, the founder of Ericsson Telephone, had in his automobile a traditional land-line telephone terminal, which he would attach to available telephone wires with a pair of long poles in order to make calls when he was away from his office (Farley 2003).[11] In the wake of the *Titanic* disaster (1912), radio was established as a standard part of ocean transport. In the 1920s, the Detroit Police Department equipped some of its patrol cars with radios (Manning 1996; Dobsen 2003).

After World War II, AT&T commercialized a mobile radio system, first in St. Louis (Farley 2005). This kind of system proved to be popular among contactors and engineers who needed to travel to distant building sites. In New York in 1976, when there was capacity for about 550 subscribers, there was a waiting list of 3,700 potential customers.

The early radio-based systems were not what we would call mobile cellular telephony. There was not, for example, the "handoff" between different cells as the individual moved from the area covered by one base station to another. Thus, the systems were not able to cover large areas. The idea of a cellular system for mobile communication was developed by D. H. King of Bell Labs in 1947 (Farley 2003).

True cellular mobile telephony was first realized in 1969 on the Metroliner trains between Washington and New York. Later, the first hand-held device was developed by Martin Cooper at Motorola. In the 1970s the system gained more and more subscribers.[12] At the same time, there was the development of a host of mutually incompatible systems in various countries that went under such acronyms as AMPS, NAMPS, PAMTS, TDMA, and TACS (Farley 2005). The Nordic Mobile Telephone (NMT) system was the first system to allow international roaming, albeit at first only within the Nordic countries. Based on this system, the European Telecommunications Standards Institute started with the development of the Groupe Speciale Mobile (GSM) (Lindmark 2002; Farley 2005). GSM (now standing for Global System for Mobile) is currently the predominant form of mobile communication in the world. In addition, there was the development of the CDMA system in the United States and the technically related iMode system in Japan.

▪

Mobile communication, and specifically mobile telephony, has a relatively long technical history. But it has been a prominent part of our social consciousness only since the mid 1990s. For example, in 1997 only a few Norwegian teens had a mobile phone. Two years later, after the introduction of pre-paid subscriptions, the subsidizing of mobile handsets, and the "discovery" of SMS by teens, more than 70 percent of 16–17-year-olds had one. In 2001 the percentage was 90, and by 2005 it was difficult to find a Norwegian teen without a mobile phone. Thus, in less than 10 years the situation went from very few teens' having a device to its being a ubiquitous part of teens' daily life.

According to the International Telecommunications Union, in 2005 there was approximately one mobile phone subscription for every third person in the world (ITU 2005). By way of comparison, about half as many persons had access to the internet.[13] Europe was the continent with the highest subscription rate (about 82 per 100 persons). Oceania was next, with 69 per 100, then the Americas, with 52 per 100. Asia had 22 per 100 persons; Africa had 11.

According to the ITU, as of 2005 there were 19 countries in which the number of mobile phone subscriptions exceeded to the number of persons (of all ages) living in the country. The most "over-subscribed" country was Luxembourg, with nearly 1.40 mobile phone subscriptions per person.[14] Also "over-subscribed" were Norway, Denmark, Italy, Hong Kong, the United Kingdom, Lithuania, and Estonia. At the other end of the scale, with less than one mobile phone per 100 persons, were Nepal, the Democratic Republic of Congo, New Guinea, and Ethiopia. The United States was middling in this category, with 67 mobile phone subscriptions per 100 residents.

China, the United States, Russia, Japan, Brazil, Germany, and India were, in that order, the countries with the largest absolute numbers of subscriptions. In 2005, China had 393 million subscriptions and the United States 202 million. This ranking is, however, more of a product of the large populations than of diffusion of the technology. Indeed, several of these countries ranked far down the list in percent of the population owning mobile phones. India, for example, ranked seventh in absolute number of mobile phone subscriptions, but there were only 7 mobile phone subscriptions per 100 persons. Thus, although there were a large number of phones in India, that country was near the bottom of the list (number 162 of 206 countries) in percent of persons owning phones.[15] Only Germany (with 96 subscriptions per 100 persons) was among the highest-ranking countries in both absolute number of subscriptions and number of subscriptions per 100 residents.

■

We increasingly use mobile phones to navigate through the day and to coordinate our activities. Teens are using the mobile phone to exchange jokes, to keep in contact with one another, and to plan their days. Preteens are using the device to maintain contact with their parents and as a vicarious umbilical cord. Young adults are using mobile communication in their pursuit of careers and in the establishment of more or less stable relationships. Parents are using it to decide who should pick up the children and who should go shopping for groceries.

Clearly one of the main uses of the mobile phone is voice interaction. In spite of all the discussion regarding new digital services and internet access via wireless devices, voice interaction is the core function of the mobile phone. In this capacity, mobile communication is challenging the traditional land-line system. In Norway in 2004, more that 66 percent of the calls were to mobile phones (PT 2004). In the United States, where rate systems and tradition favor voice-based interaction, about half of all calls last less than a minute and 70 percent last 2 minutes or less (FCC 2005).

Another mode of mobile communication is text messaging by means of the Short Message System (Trosby 2004). This form of interaction is one of the surprising innovations associated with mobile telephony. It was not even available until 1993. In 2005, according to GSM World (2007) and Logica CMG (2006), at least a trillion text messages were sent.[16] If we divide a trillion messages by the 2.1 billion mobile phone subscriptions that were active in the world in 2005, each subscription would have had to send about 470 messages per year—about 1.3 per day. Obviously not all people use text messaging, and there are cultural and socio-demographic differences in use of the medium. In Norway, an average mobile phone user sends more than five text messages per day, and in 2006 teenage girls sent a mean of 23 per day.[17] Users in the Philippines, one of the most SMS-oriented countries in the world, send about seven messages per day (National Statistical Coordination Board 2007).

▪

Beyond being a device with which we carry out functional activities, the mobile telephone has emblematic status in modern society (Fortunati et al. 2003). Simply knowing what kind of mobile phone someone owns provides some of us with insight into that person's social position and status. We recognize a "teeny-bopper" with a Nokia, or an advanced business user with a Trio, a BlackBerry, or a "Smart Phone."

The very form of the mobile telephone has become a minor cultural icon. A child can buy a helium-filled balloon in the shape of a mobile phone. Thus the mobile phone competes with the forms of Pokémon and various Disney characters in the marketplace of children's iconography. All of this points to the notion that the mobile phone is an object that has become symbolically located in our sense of culture and identity.

The device itself can ring, peep, squawk, or play a Coltrane riff at the most unexpected moment. In doing this, it rearranges the co-present interaction and demands that we redirect our attention into other realms. Its use in co-present situations is symbolically invested and, as I will examine below, can be seen in terms of its contribution to ritual interaction.

The mobile telephone presents us with a Garfinklian breaching experiment writ large (Woolgar 2003). Goffman examined the social order when he examined how it was challenged by mental patients, and there is some of the same in looking at our adoption and use of mobile communication, albeit at a more mundane level. By examining how the existing order has been challenged by the introduction of a new and ever more insistent device, we can examine the functioning of society.

Plan of the Book

In chapters 2–6, I trace the intellectual history of ritual interaction and its contribution to social cohesion through Durkheim, Goffman, and Collins. In chapters 7–10, I apply this to mobile communication.

In chapter 2, I examine the idea that there is a loss of social cohesion in society. On the one hand, there are questions as to the viability of social cohesion and the similar concept of social capital. The history of social capital is traced through its development by Bourdieu, Coleman, and Putnam. This is then contrasted with the sense that there is increasing individualization in society (Beck and Beck-Gernsheim 2002). Thus, the social fabric is seemingly fraying and at the same time, we are experiencing a shift from Tönnies's more traditionally oriented Gemeinschaft to his notion of the more contractual and individualized Gesellschaft.

The introduction of information and communication technology has become a part of the discussion on this point. In the balancing between sociation and individualization, the question becomes "What is the role of the various new forms of technology?" Putnam (2000) lays some of the blame for the loss of social capital on television. The radio (Douglas 1999), the traditional land-line telephone (Fischer 1992; Rakow 1992), the personal computer (Kraut et al. 1998), the internet (Nie 2001; Kazmer and Haythornthwaite 2001; Quan-Haase and Wellman 2002), and the mobile telephone have also figured in the tension between the collective and individual impulses of society.

In chapter 3, I examine the work of Durkheim as it relates to ritual interaction. Indeed, the social analysis of ritual is often traced to the work of Durkheim. In *The Elementary Forms of Religious Life* (1995) he lays out the way in which solidarity and shared symbolism is developed in the context of group interaction. Implicit in this more specific analysis is the more basic Durkheimian question "What holds society together?" According to Durkheim, it is ritual.[18] Society is held together through stronger and weaker and more or less elaborate rituals. The solidarity provided by rituals precedes that provided by more utilitarian forms of interaction such as economic and functional dealings. Ritual is the link between group structure and group ideas.

In chapter 4, I examine how Goffman reapplied Durkheim's notions in a scaled-down version. Goffman encourages us to apply Durkheim's notions to everyday situations. In addition, rather than looking at individuals, Goffman focused on situations as his basic unit of analysis. These might be overt presentations of a constructed facade or more "back-stage" situations in which the facade is dropped to a greater or a lesser degree.

Collins's notion of interaction ritual chains (chapter 5) helps to integrate the work of Durkheim and Goffman into a model that can be applied to commonplace situations. The model helps us to understand the roles of both large-scale and small-scale interactions in the ongoing process of daily life. It is a lens through which we can see how seemingly mundane interactions are indeed the small rituals with which we celebrate our connection to the broader sphere. It is through the rehearsal of these interactions that we practice Goffman's deference and demeanor. It is though these micro-interactions that the significance of artifacts is revitalized, and it is through our use of these interactions that we recognize the individual as a significant element in focused ritual interaction. Collins also shows how the effervescence of larger events is an important aspect of social cohesion.

The material of chapters 3–5 is brought together in chapter 6, where I examine how ritual interaction is a catalyst for the development of social cohesion. Ritual leads to cohesion through a kind of syllogism. For example, if I attend a sporting event, a revival meeting, or a dinner party, and I become engrossed (or "entrained," to use Collins's term) by the situation, as does the fellow who is next to me, that is not enough to engender solidarity. In order for the situation to be seen as a ritual and thus capable of engendering solidarity, we must mutually realize that we are both entrained. I have to see that he is also drawn into the events, and he has to see that I am likewise engaged. When we arrive at that mutual understanding and when we establish a common sense of mood, it breaks down the barrier between us and provides us with a common experience upon which we can start to build our interpersonal interaction. It is in this way that the ritual situation is a catalyst that can result in the mutual recognition of engagement. The product of this is some form of solidarity. Goffman helped us to see that these situations can be quite simple and fleeting. The point is, however, that there is a mutually recognized situation and the establishment of a common mood.

I apply this analysis framework to mobile communication in chapters 7 and 8. In chapter 7, I examine how mobile communication affects co-located interaction, and in chapter 8, I examine ritual interaction in mediated situations. The material here shows that the mobile telephone is sometimes a disturbing element in co-present interaction, but also that it augments the possibilities for cohesion. I examine this finding in terms of greeting sequences, the negotiation of romantic involvements, texting argot, repartee, jokes, and gossip. The general finding here is that mobile communication supports the development of cohesion by expanding the flow of interaction beyond face-to-face meetings. I do not claim that the mobile phone is alone responsible for the establishment of cohesion, but I argue that it plays into the enhancement of social cohesion in small groups. In chapters 9 and 10, I look at other research that supports this finding. There is a growing consensus that mobile communication supports the concretion of small groups, particularly family and friends. Work done in Korea, in Japan, in the United States, and in Europe supports this general finding. In chapter 9, I examine how Licoppe's concept of connected presence helps us to understand this general direction of social interaction. Finally, in chapter 10, I examine some of the other issues associated with the tightening of small group interaction that has seemingly been supported by mobile communication, including the development and strengthening of local ideologies and the tendencies that support but also mitigate the effects of over-configured social ties.

Comments on the Method

In this book, I draw on several types of empirical information when discussing the use of ritual in mobile interaction. These include quantitative studies that have been carried out around the world. In addition, I include qualitative analyses of mobile phone use. The qualitative analyses include both interview situations and, in the spirit of Goffman, observations. It was the suggestion of Emmanuel Schegloff to pursue the use of mobile communication in natural settings that was the motivation for including observational approaches.[19] Schegloff correctly suggested that the steering of questions in quantitative data collection and in qualitative interview situations can bias the outcome of research. Thus, his encouragement was to adopt a system of natural observations when investigating the use of mobile communication. This direction in research is also Goffmanian. Goffman encouraged researchers to pay attention to the minor "grunts and groans" of their subjects (Goffman and Loflund 1989). He noted that researchers must force themselves to be a witness to what is happening around them. This is a difficult skill, since we are generally taught not to be over-concerned with the affairs of others around us. A researcher, and particularly a researcher using observational techniques, must set aside some of these fine points and focus on the observation of others (Goffman and Verhoeven 1993, p. 327). In the case of the work presented here, this took several forms. At the most reserved level, I simply observed the use of mobile telephones in various public settings. The strategy was to position myself in some public place where I, in essence, had a legitimate reason be there (a café, a tram, a bus station, in the park, along the street, etc.). The point was to simply wait for people to begin using their phones. When I saw someone using one, I focused on the situation. I tried to keep track of how the individual dealt with the telephone conversation; at the same time, I tried to watch how he or she interacted with others who were there. I tried to "note their gestureal, visual, bodily responses to what is going on around them" (Goffman and Loflund 1989). How did they approach or avoid each other? What special care was given to the telephonist, and how did the telephonist negotiate the balance between the co-located people and the person to whom he or she was speaking?[20] At this stage in the work, I was not interested in the dialogue of the individuals; rather, I was interested in how they carried themselves in the situation. In addition, I was interested to see whether others reacted to their use of the device. Did they provide visual or verbal cues to the individual that the use of the device was not appropriate? Did they avoid the telephonist, or did they accommodate the situation?

The next (relatively minor) phase was to carry out several small experiments with co-location. In these experiments, I would find a person talking on a phone and then move within a meter of that person. In a grocery store, for example, I would move next to the telephonist and start to look at the labels of the tomato cans on the shelf. In a train station, if the telephonist was sitting on a bench, I would sit down nearby. I was not interested in confronting the telephonist by looking him or her in the eye. I was only interested in moving into the periphery of the person's intimate sphere (approximately within arms' reach). The point of these "gonzo" experiments was to find the degree to which the telephonist was cognizant of others in his or her co-present location. Was the telephonic interaction so engrossing that the telephonist would simply continue without giving way to other types of activities occurring nearby?

The final observational technique was to move into audible range—the range of Goffman's grunts and groans—of ongoing mobile phone conversations. I tried to overhear the conversations of the individuals. I took to heart Schegloff's suggestion to focus on naturally occurring situations and Goffman's idea that the person in the field should try to come close into the situation of the individuals and to not be afraid. I wanted to gain access, to the degree that it was possible, to the flow of events in the individuals' lives. Once again, the point was to be "tuned into something that you then pick up as a witness—not as an interviewer, not as a listener, but as a witness to how they react to what gets done to and around them" (Goffman and Loflund 1989). Indeed, in the same piece, Goffman says that the researcher has to be open to "being snubbed." By playing with that boundary, the researcher is moving in as close as possible and gathering a rich picture of the situation.

As it turns out, it is quite easy to tune into mobile phone conversations in public spaces. People often position themselves in a static location (a bus seat, leaning against a wall, standing in a waiting area, etc.). I was able to blend into the scene and take on the appearance of a fellow shopper, passenger, or the like. Again, I would not make visual contact with the telephonist or try to talk to him or her. I would only listen and observe. I would then commit as much of the interaction to paper as I could, often within a few minutes.

My field notes were just that: handwritten notes that often give only a sketchy outline of an event. These were then developed into entries that were more complete and then organized in terms of the analysis.

I made little use of photography or voice recording, because it is harder to blend into a scene when using a camera or a recording device.[21] From my experience, a researcher is more intrusive when using photo or video equipment. The positive side is that, when successful, one has a rich record of the event. The down side is that one is more obvious and perhaps disruptive. The alternative is to observe manageable units of social interaction, to practice learning to read situations, and to see the significant elements and note them as soon and as fully as one can.

There is also the ethical question of using material provided by others without their permission and indeed without their knowledge. I overheard one side of someone's telephone conversation, and I made assumptions about that person's life and situation. These assumptions might have been incorrect, and I might have heard things to which they would rather I not be party. A slight "fig leaf" is that the individuals were broadcasting their lives to anybody who took the time to listen. I never sought out situations where there was a reasonable chance that the telephonist would think that he or she was in a private sphere. All the observations were made in public areas where there were many other bystanders. Thus, it was abundantly clear to the telephonists that we were in open settings. Another "fig leaf" is that I never heard the names of the individuals involved. It would be impossible for me to find my way back to them and confront them with their utterances.

Conclusion

The entrance of the plumber was, from my perspective, an example of how mobile communication has interrupted the flow of normal co-present ritual interaction. There was not the normal greeting that we expect when receiving visitors into our home. There was no salutation, no handshake, no exchange of names. This was a breach. I experienced the situation as an example of flawed etiquette. It was a failed ritual.

From the perspective of the plumber and his interlocutor, however, the telephone call allowed them to maintain and develop their ongoing relationship. The tone and the decorum were appropriate. Turn taking was probably fluid and allowed each speaker the space he or she needed to make comments and replies. The greetings and the leave taking probably were carried off without a hitch. The two might have shared some private jokes or exchanged a few speculative comments about other friends. If the other person was the plumber's boss, for example, he may have given the plumber some orders on how to deal with certain customers or suppliers, and they may have engaged in working out the power relations between them. In other words, the plumber and his interlocutor were engaged in a mediated ritual interaction that extended their pre-existing interaction.

The entrance of the plumber shows how mobile communication is being played out in society. It can result in awkward co-present situations, but it can support better contact within the personal sphere. It is causing us to rethink co-present ritual interaction, but it is also extending our interaction with friends, family, and colleagues beyond simple co-presence.

2
ICT and Tension between Social and Individual Impulses

It may be a bit unexpected to start a book on mobile telephony by going back to the dawn of sociology. After all, what do those "dead Germans" (as we called them in our graduate sociology seminars) have to do with information and communication technology (ICT)? The answer is that in many respects the older generation of sociologists was dealing with the question of social cohesion in the face of technological development.

Durkheim, Marx, Simmel, and Weber were sifting through the consequences of industrialization, which gave European society a thorough reshuffling. Almost no institution was left untouched. The nature of work and the idea of the corporation of course were fundamentally changed. When these early social thinkers were working, there had been the rise of the bureaucracy. Work had increasingly become factory-based wage labor. The authority of the church was reduced. The family went from being extended and often locally based to being nuclear and far more mobile. The cities became centers of production and trade. In short, not much was left of the old ways.

This, of course, was fertile ground for the social scientists. There was Tönnies's Gemeinschaft and Gesellschaft (1965), Marx's class-in-itself and class-for-itself (1995), Durkheim's mechanical and organic solidarity (1954), and Weber's rationalism (2002). These early theorists were trying to describe the effects of the industrial revolution on society. In each case, there was an exploration of the tension between the collective and the individual. This has elements of the present-day discussion regarding social capital or communities of practice and individualization. The discussion of the forefathers and the current discussion are similar in this way (Beck and Beck-Gernsheim 2002; Boase et al. 2006; Bugeja 2005; Katz and Rice 2002; Kavanaugh and Patterson 2001; Kraut et al. 1998;

Nie et al. 2002; McIntosh and Harwood 2002; McPherson et al. 2006; Putnam 2000; Wolak 2003).[1] There are, however, differences. Whereas industrialization was the motivating force in the historical discussion, the mobile telephone and the rise of the internet are more central to the current discussion. In addition, while ICTs have changed many things, they do not seem to have had the same fundamental influence on the broader institutions as did industrialization. For example, approximately half a century after the invention of the transistor, the cities are largely though not completely unchanged. The church has not regained its former central position in society, wage labor and the nuclear family are about the same, and so on. The same could not be said for these institutions 50 years after the onset of industrialization.

Although the institutions were not as changed by the rise of ICT as they were by the rise of steam-based industrialization, there is a similar discussion regarding the tension between the collective and the social. In the internet revolution, social capital has become the vehicle for a portion of this discussion.

Social Capital and Cohesion

The question of social capital—and its opposite twin, individualism—goes to the core of sociology.[2] It is very often in this context that social cohesion is discussed. It raises the question of the "impulse to sociability" as discussed by Simmel (1949) vs. the striving toward individualism as seen in the discussions of Lash, Giddens, and most particularly Beck. (See Beck et al. 1994.) It is indeed this issue that motivates my work on mobile telephony.

At the most abstract level, how is it going for social cohesion? Is the joinery of society holding up when all the bits seemingly want to go their own way? Thinking more specifically about mobile communication, we can also pose the same questions as do those who study the internet and television: Does mobile communication contribute to or destroy social cohesion? Is this technology more advantageous for the intimate sphere? Alternatively, does it favor broader contacts with the so-called weak links?

Perhaps motivated by the work of Robert Putnam, particularly his book *Bowling Alone* (2000), the concept of social capital has been applied to a wide range of situations, including the potential for Third World development (Fox 1996), community governance (Bowles and Gintis 2000), social and economic outcomes (Woolcock 2001), self-rating of health (Kawachi et al. 1999), and children's math skills (Morgan and Sorensen 1999).

The concept of social capital, of course, has arisen against a backdrop of an apparent fraying of the social fabric brought on by the adoption and use of technologies such as the mobile phone and the internet. The analysis of social capital brings us back to the broader question regarding the status of the social cohesion as opposed to the individual. It includes the implicit recognition of social networks and the role of social ties, and the news from this front is worrying. According to McPherson et al. (2006), commenting on the scene in the United States, the number of confidants with whom respondents discuss important matters fell from 2.94 in 1985 to 2.08 in 2004. McPherson et al. note that attrition has been seen in both kin and non-kin confidants. In this respect, Portes's (1998) observation that the analysis of social capital is at its core an extension of the traditional social project of tracing social cohesion is on the mark.

In early analyses of social capital (Bourdieu 1985; Coleman 1988) there was no explicit examination of the concept with respect to ICTs. Recent analyses of social capital, however, are more unequivocal in this connection. While there are some questions as to the influence of the personal computer and the internet, as we will see in chapter 9, there does seem to be covariance between mobile phone use and informal social interaction (Ling et al. 2003).

Social capital as conceived by Bourdieu is seen as a way to distinguish between characteristics of the individual—economic and cultural capital—and those of the social group. He adopts the concept of capital as a kind of "stock" that is built up or dissipated over time. Instead of seeing the outcome of situations as merely a zero sum, this notion of capital asks us to look at the elements that are carried across the different transactions/situations. Bourdieu famously suggests three forms of capital that include economic capital (the traditional use of the concept), cultural capital (the individual's sense of taste and refinement, where we develop a sense of cultural insight), and social capital.[3]

Economic capital is associated with personal ownership accumulated over time and "has the potential for produce profits, to reproduce itself in identical or expanded form" (Bourdieu 1985, p. 244). Further, Bourdieu states, economic capital is that "which is immediately and directly convertible into money and may be institutionalized in the form of property rights" (ibid., p. 244). This is, of course, the most common definition of 'capital', as in the term 'capitalist'.

With social capital, the focus is on the interpersonal and social resources that a person can contribute to as well as draw on in everyday life. It is perhaps best seen as a kind of metaphorical common bank account where the individual can make various types of contributions. We can help our neighbors change the tires on their car, make them some chicken soup when they are sick, or help them raise a barn. At a more abstract level, we can contribute to famine relief or buy Girl Scout cookies. In each case, we are paying into the "social capital" account. In return, we can perhaps receive help fixing a bicycle from the neighbors or ask them to pick up some items at the store on another occasion. We might find that the Girl Scout from whom we bought the cookies might be able to help out with baby-sitting.

The concept of social capital allows us to focus the fungibility of social interactions. It argues against a simpler exchange or market-based metaphor. It argues against the notion of society based on "a discontinuous series of instantaneous mechanical equilibria between agents who are treated as interchangeable particles" (Bourdieu 1985, p. 241). To use Bourdieu's image, to see society in this way is to see it is as a game of roulette in which fortunes, memories, culture, etc. can be won or lost at the spin of the wheel. Nothing is stable; everything is in play with each interaction. Social capital argues that we should look at the residue and not just the flow.

Social capital is not the possession of an individual but of a collective. It is a characteristic of the social situation in which the individual finds him or herself. As individuals, we can contribute to the development of social capital by helping to maintain the group. An individual can bake a cake for the Friday get-together, tell a joke, or help a neighbor rake leaves. An individual can also draw on the social capital of the group. Perhaps one can expect another member of the group to remember to bring soda when the "Friday cake" is baked, or to laugh at a joke. Perhaps a neighbor will volunteer help when we are sick. In addition, one can draw on the group's social capital when, for example, one represents the group and speaks in its name with the backing of its collectively owned capital. It is easy to see how mobile communication that is more individually based and more omnipresent than traditional land-line telephony would influence this conception (Hjorth 2006).

▪

Building on Bourdieu's work, the sociologist James Coleman contributed to the understanding of social capital. He considered the mechanisms that generate social capital and the consequences of situations in which social capital has been generated. In particular, he introduced the concepts of reciprocity, trust, and gifting, which are important aspects of social capital (Coleman 1988).[4]

Coleman examines the way that closure within the social network affects social capital. There is the idea that the multiplicity of ties has the consequence of ensuring norm observation. It is in this context that he introduces, perhaps one of the most frequently cited illustrations of social capital in action, namely the interaction of diamond cutters in New York City's diamond district on 47th Street.[5] It is common, for example, for one diamond merchant to trust another with a collection of uncut and therefore untraceable stones worth a large sum of money. There is no need for collateral or other guarantees, since any malfeasance would result in exclusion of the individual from the community of diamond dealers. Coleman describes how this exclusion is not only from a particular business but from a whole set of social connections.

Coleman examines how their business interactions are prefaced on a trust that has been built up via a collection of other intersecting factors. Since the diamond trade is largely Jewish, there are familial, religious, and ethnic ties that hold the community together. There are links of a common cultural background, intermarriage, membership in the same civic institutions, etc. In addition, there are links based on propinquity since many of the dealers live in the same community in Brooklyn. Exclusion therefore means not simply the loss of economic position, but also shunning. The weight of the social links forces a kind of respect and coherence within the community at the risk of exclusion.

The advantage of this well of trust is that it facilitates interactions. There is not the need for cumbersome guarantees and contractual negotiations. You simply know that another member of the community is to be trusted, and that is enough.

Coleman also draws our attention to the idea that social capital is necessarily a social construction. In order to understand this he develops a thought experiment in which a person is offered a better job in another community. Since this would require moving, it means that the decision to take or not to take the job also has consequences for the local community. If the person decides to uproot the family and move, the social dynamics of the community of origin are disturbed. It might be that he or she is the local scout leader, a member of the local fire brigade, or the best soprano in a church's choir. The move would mean that there was a small hole in the social fabric. In addition, it is not necessarily such that the individual would be able to make the same attachments quickly in the new community. The experiment is not a plea for stasis, but it does serve to illustrate the social nature of this kind of capital.

Another aspect of social capital illustrated in the above-mentioned thought experiment is that often social capital is a by-product of other activities. The individual's interaction with others in the role of being a scout leader, a member of the fire crew, or the best soprano in the church choir is not simply the specific task at hand, but also the other connections and associations that come from the interaction. There was the time that the scout master drove one of the youngsters home and got a cup of coffee in exchange. There was the camaraderie of with the others on the fire crew and the fun they had at the county-wide conference of firefighters, or the time the choir was asked to sing at a wedding and the leader forgot the sheet music. It is in these interactions that cohesion is developed and maintained. Coleman (1988, p. 118) writes:

The public good quality of most social capital means that it is in a fundamentally different position with respect to purposive action that are other forms of capital. It is an important resource for individual and may greatly affect their ability to act and their perceived quality of life. They have the capability of bringing it into being. Yet, because the benefits of actions that bring social capital into being are largely experienced by persons other than the actor, it is often not in his interest to bring it into being. The result is that most forms of social capital are created or destroyed as by-products of other activities. Social capital arises or disappears without anyone willing it into our out of being and is thus even less recognized and taken into account in social action than its already intangible character would warrant.[6]

Coleman developed the idea of social capital by bringing up the notions of closure in the group and also the sense that it is a social resource. We can look to Mark Granovetter and Ronald Burt for the notion that openness in the social group is also an important issue when thinking of social capital.

.

While there is the sense that trust and reciprocity (that is, social cohesion) arises in tightly bound groups, there is also a paradoxical idea that openness is also needed in social groupings. On the one hand, Coleman's work suggests that it is in tightly bound, multiplexed groups that we are most likely to find reservoirs of social capital. It is in these groups that we find the highest degree of self-disclosure. There is the notion of confidence in other's ability to pitch in and help when needed. There is the sense that "we are all in this together."

But if society were only made up of tightly bonded groups, the flow of information and resources would be hindered. If society were divided up into fully configured and multiplexed groups as described by Coleman, it might well be the case that there would be enormous trust within the group, but suspicion and mistrust between groups. It would be difficult to develop interactions across cliques or clans. There would be, for example, little opportunity for the development of enterprise across the groups. It would be difficult to find jobs when there was the need for more workers and it would be difficult to develop romantic relations across the groups. In addition to the functional boundaries between the groups, there may be ideological frictions, a kind of us-vs.-them situation. Indeed, this dynamic moves the action in the play *Romeo and Juliet*. According to Putnam, this hypothetical society would be "balkanized."[7] The rise of mobile communication may be pushing society in this direction to some degree.

The person who has studied the role of weak ties and their role in society most carefully is Mark Granovetter. He has examined these "bridges" or weak ties and the way that they provide for the efficient diffusion of information (1973). He shows how weak ties between groups allow for the efficient flow of information. The use of casual connections, as opposed to the more heavyweight interpersonal links described in Coleman's work on diamond cutters, are more spurious and are not necessarily encumbered with the same levels of trust and reciprocity. But "whatever is to be diffused can reach a larger number of people, and traverse greater distance (i.e. path length), when passed through weak ties rather than strong ties" (ibid., p. 1336).

Since all the members of a clique share the same information, there is little chance that new opportunities will be discovered there. Thus, innovations are less likely to be spread internally. In the same way, groups with few weak ties will be isolated in this sense. Granovetter (ibid., p. 1369) describes research by Rapport and Horvath that examined the functioning of a social network:

> They asked each individual in a Michigan junior high school ($n = 851$) to list his eight best friends in order of preference. Then taking a number of random samples from the group . . . they traced out, for each sample, and averaged out of all the samples, the total number of people reached by following along the network of first and second choices. That is, the first and second choices of each sample member were tabulated, then the first and second choices of these people were added in, etc., counting, at each remove only names not previously chosen, and continuing until no new people were reached. The same procedure was followed using second and third choices, third and fourth, etc. up to seventh and eighth.
>
> The smallest total number of people were reached through the networks generated by the first and second choices—presumably the strongest ties—and the largest number through the seventh and eighth choices. This corresponds with my assertion that more people can be reached through weak ties. A parameter in their mathematical model of a sociogram, designed to measure, approximately, the overlap of acquaintance circles, declined monotonically with increasing rank of their friends.

It is through the weak ties, or at least the less sought after ties, that individuals could be reached with new information, ideas, and inspiration. The same point is made in Fine's analysis of Little League baseball. Fine (1987, p. 180) found that Little League players, through weak networks of persons who they met at camp or in other milieus outside the more routinized baseball team, were able to refresh their reservoir of profanities, racial epithets, and off-color jokes. After all, in their strong-tie groups, the boys were basically sharing the same jokes and profanities. There was little creative innovation. It is through the weak or bridging channels that the new influences were brought into the core group, for better or worse.

Another application of Granovetter to activities in the commercial world has been done by Burt, who has examined the role of weak ties as being information entrepreneurs (Burt 1999, 2000, 2001b; Burt et al. 1998). Following this line of thought, persons who stand on the boundary of a group but maintain ties with others groups are an important resource. They are able to observe innovations and activities in the adjoining social clusters. Based on their ability to recognize novel developments, they can even be assigned the special role of innovator in their primary group. Burt also examines how if a particular link results in many innovations, others in the primary group may also seek out links from the same milieu in order to cultivate innovations.

Granovetter speculates about the interaction between the weak and the strong links. He asks, for example, if a person gets information on a new job (or for that matter, a new boyfriend or girlfriend, a where a new car can be found, the location of a party, etc.) through a weak link, there may be the need to legitimize the choice through the strong ties. There may be the need to gather information as to the authenticity of the information, the desirability of acting on it, the degree to which the source can be trusted, etc. The ability of the "strong tie" group to endorse the opportunity can play in to whether the individual will move ahead with the opportunity.

There is also a balancing here. On the one hand, the individual needs to have new ideas and opportunities. However, if the primary group lays a blanket on all such activities the individual may feel hampered. Following from the work of Gergen, there may be the development of a local group ideology that makes the alternative ideas impossible (Gergen 2003, 2008). This tension is seen in *Romeo and Juliet*. The initial weak tie between the "star-crossed lovers" would not have been played out had they accepted the group ideology of bad blood between of the Montagues and the Capulets. The ideologies of the cliques augured against the weak ties and limited interaction. It is, of course, this tension that gives the play its power.

Tension in the balance between the clique and the broader group may be one of the influences of the mobile telephone. That is, the device lowers the threshold for interaction within the peer group to the degree that it readjusts the ratio between strong and weak ties.

■

Robert Putnam has, perhaps more than others, put social capital on the map. He has examined the issue and applied it to historical material in the United States. His work has made social capital a popularly recognized concept.

In a sense, Putnam had different intentions than Bourdieu or Coleman. Rather than work on the development of social capital as a theoretical concept, he was more concerned with applying it. Putnam operationalized social capital and drew on a broad range of material in order to examine the ebbs and flows of social capital in the United States. In *Bowling Alone* (2000) he examines the participation trends in politics, civic interaction, religion, at work, in informal settings, and in volunteer and altruistic groups. In almost every case, Putnam reports that social capital is on the decline in the United States. In summary, he writes:

The dominant theme is simple: For the first two-thirds of the twentieth century a powerful tide bore Americans into ever deeper engagement in the life of their communities, but a few decades ago—silently without warning—the tide reversed and we were overtaken by a treacherous rip current. Without at first noticing, we have been pulled apart from one another and from our communities over the last third of the century. (ibid., p. 24)

Putnam identifies four major reasons for the decline in social capital: television, suburbanization, increased working (meaning less time for socializing), and the passing of the pre-World War II generation. The first two, and the third to a degree, reverse Simmel's "intensification of emotional life" (1971, p. 324). Putnam outlines how TV, rather than engaging the individual in social interaction, steals time that might otherwise be spent in interaction with neighbors or in other civic activities. As an extension of this, there has been the development of various electronically based activities that individualize leisure time. Surfing the internet (though perhaps not chatting, web-based gaming, or using email) and the use of gaming machines such as Nintendo, Xbox, Wii, and Game Boy also take people away from social interaction into individualized relaxation.

Television and other types of individualized entertainment may capture the attention of the individual, but suburbanization cuts us off physically from others (Gergen 2008). Having to travel greater distances to activities means that visits become a commitment that cannot happen casually. An automobile trip is often necessary before we can participate in any form of either formal or informal social interaction (Hjorthol 2000). The effort required to interact means that there is little time or effort afforded to social interaction.

The final reason outlined by Putnam is that the pre-World War II generation—a generation distinguished by a strong sense of community forged by their dramatic common experience—is passing from the scene. The fact that this generation experienced the Great Depression, dove almost directly into the horror of World War II, then raised the postwar Baby Boom gave the generation a common experience and left them with the motivation to establish various civic institutions. It is this generation that established or at least gave vitality to organizations such as the Knights of Columbus, Scouting, Parent-Teacher Associations, and Putnam's bowling leagues. None of the generations that have come since has had the same life experience. Though the war in Vietnam set its mark on the postwar children, this was seemingly not as profound or as encompassing as the experience of the pre-World War II generation. The passing of this generation and the various individualizing effects of working life, television, and the suburbs are the forces that, according to Putnam, are pulling at the trusses of social capital. The passing of this generation and their replacement by the postwar Baby Boomers, "Generation X," and their successors means that a large reservoir of social capital is also passing from the scene, according to Putnam.

Tying this back into the question of interpersonal mediation and technology, Putnam suggests that point-to-point media such as the telephone reinforced existing social ties. Drawing on work by Fischer (1992), Putnam notes that the telephone was seen as an important social link for rural women as early as the 1930s. It has the ability to offset some of the isolation resulting from television and the suburbs (Putnam 2000, p. 168).

The result of Putnam's work is an increasing focus on the role of civic engagement in society. Putnam has been criticized for selectively using material to support his case while ignoring other trends in society. Costa and Kahn (2003) assert that Putnam understates the impact of women's increasing participation in the workforce. Their analysis weighs this factor differently. In their estimation it has led to some of Putnam's reduction in social capital. Hall (1999) finds that social capital has been stable in the United Kingdom, though Grenier and Wright's (2001) analysis of Hall's work suggests that the stability is largely found in the middle and upper classes and not among the working classes.

Fischer (2001) criticizes Putnam, but more generally he criticizes the idea of social capital. This serves as a nice transition to the discussion of individualism and its impact on society. In many ways, individualism is a mirror image of social capital. Whereas social capital examines the urge for sociation, individualism is the opposite. Whereas social capital examines processes whereby common meanings are developed and mutated, individualism witnesses their elimination and marginalization.

Individualization

In the previous section, I noted that there is a tendency toward collective behavior—in this case called social capital—but that in the eyes of some it is shifting and loosening. The opposite of social cohesion is individualism, and this, of course, is a strong current in society. Simmel's "impulse to sociability" (1949) finds its opposite in the striving toward individualism (Beck et al. 1994). It may seem that the impulse animating Simmel's sociation is being outflanked by the economically rational individual. Bauman states that "individualization is here to stay" (2001, p. 50), Matsuda asserts that individualism is a central characteristic of the scene in Japan (2005, p. 130), and Sennett describes the impact of the individual's dynamic working situations on the ability to put down roots (1998, p. 87; see also Aakvaag 2006).[8]

Scott Lash takes up the issue of individualization and suggests that we are moving into a decidedly new era. Describing the shift from Gemeinschaft to Gesellschaft, Lash writes:

The motor of social change in this process is individualization. In this context Gesellschaft or simple modernity is modern in the sense that individualization has largely broken down the old traditional structures—extended family, church, village community—of the Gemeinschaft. It has not yet fully modern because the individualization process has only gone part way and a new set of Gesellschaftlich structures—trade unions, welfare state, government bureaucracy, formalized Taylorist shop floor rules, class itself as a structure—has taken the place of traditional structures. Full modernization takes place only when further individualization also sets agency free from even these (simply) modern social structures. (1994, p. 114)

Lash suggests that a social force similar to but in opposition to Simmel's sociation has been unleashed. He suggests that the result of modernization is that communal society has been broken down and replaced by "a set of atomized individuals (ibid., p. 114). If there was a slightly gloomy sense to the notion that social capital is on the decline, then there is a clear pessimism in the work of those describing the rise of individualism. Ulrich Beck writes:

. . . collective and group specific sources of meaning (for instance, class consciousness or faith in progress) in industrial society culture are suffering from exhaustion, break-up and disenchantment. These had supported Western democracies and economic societies well into the twentieth century and their loss leads to the imposition of all definition effort on the individuals; that is what the concept of the 'individualism process' means. Yet individualization now has a rather different meaning. The difference to Georg Simmel, Emile Durkheim and Max Weber, who theoretically shaped this process and illuminated it in various stages early in the twentieth century, lies in the fact that today people are not being 'released' from feudal and religious-transcendental certainties into the world of industrial society, but rather from industrial society into the turbulence of the global risk society. They are being expected to live with a broad variety of different, mutually contradictory, global and personal risks. (1994, p. 7)

Institutions that could help to moderate the effects of life on the individual are being systematically washed away. The individual, to a greater and greater degree, has to patch together his or her own sense of self. Once you could call yourself a Midwesterner, a working-class child, or a city kid; now such forms of orientation are being replaced by a patched-together sense of self. Beck—who witnessed the fall of the Berlin Wall and the reunification of Germany—says that opportunities, threats, and the "ambivalences of biography" once provided by the family, the village, or the social class are increasingly handled by the individual. Although the family and the class still exist, they are not the same stable institutions. The project of creating an identity has become more complex. Individuals are expected to master "risky opportunities" (dealing with the health-care system, deciding on personal finance, purchasing ever more complex products, and so on) without have an adequate background or an adequate overview (Beck 1994, p. 8). Beck adds:

'Individualization' means, first, the disembedding and, second, the re-embedding of industrial society ways of life by new ones, in which the individuals must produce, stage and cobble together their biographies themselves. Thus the name 'individualization'. Disembedding and re-embedding (in Giddens words) do not occur by chance, or individually, nor voluntarily, nor through diverse types of historical conditions, but rather all at once and under general conditions of the welfare state in developed industrial labor society, as they have developed since the 1960s in many Western countries. (ibid., p. 13)

The trend toward individualization is not necessarily driven by the desires of the person; rather, it is a condition of modernity. Beck plays on Sartre when he says that people are "condemned to individualization" (ibid., p. 14). Our biography is something that we "do ourselves." Whatever I think or do becomes my biography. "That does not necessarily have anything to do with civil courage or personality, but rather with diverging options and the compulsion to present and produce the 'bastard children' of one's own and others' decisions as a 'unity.'" (ibid., p. 15) Beck suggests that, although people "communicate in and play along with" the older forms of institutions, there is an overall withdrawal and lack of commitment. People move to other "niches" of identity and activity. In other writings (e.g. Beck and Beck-Gernsheim 2002), Beck has even discussed the oxymoronic idea of the institutionalization of individualism.

At a technical level, there also a seeming stream of technologies that atomize social life and cater to the individual as opposed to the collective. The automobile, the personal stereo system, and indeed the mobile phone are technologies of the individual. Collective solutions for transit (trams and buses), for listening to music (the concert, the piano, the phonograph—without headphones), and for interpersonal communication (the land-line telephone) have started to be replaced by technologies owned and controlled by the individual. We choose when to come and go, we choose music to fit our taste, and we choose when and where to communicate.

Mediated Interaction and the Balance between the Social and the Individual

There is still the sense that we somehow manage sociality. Lash comments that Beck does not consider the potentials for mediation of social interaction. In the foreword to Beck and Beck-Gernsheim's book *Individualization,* Lash writes:

I do not think that that the technical dimension is sufficiently taken on by the Becks. Nor the dimension and extent to which social relations are mediated through the (now interactive) mass and non-mass media of communications. Individualization, the Becks argue . . . is a question of "place polygamy." My point is that such "place-polygamy" is always necessarily technologically mediated, by cheaper air flights, by mobile phones, by microprocessors in various small boxes, by protocols and channels enabling communication at a distance. (Beck and Beck-Gernsheim 2002, p. xii)

Thus, Lash holds out the possibility that sociation is still alive and well, albeit in a different form. Further, ICTs are influencing this transition. Giddens (1994, p. 186) also comments on the new situation confronting us. His point is that while many of the structures that have stood for the production of community (associations, trade unions, local organizations, etc.) are waning, we are also able to recreate these structures by other means:

We need . . . to question today the old dichotomy between "community" and "association"—between mechanical and organic solidarity. The study of mechanisms of social solidarity remains as essential to sociology as it ever was, but the new forms of solidarity are not captured by these distinctions. For example, the creation of "intimacy" in post traditional emotional relations today is neither the Gemeinschaft nor Gesellschaft. It involves the generating of "community" in a more active sense, and community often stretched across indefinite distances of time-space. Two people keep a relationship even though they spend much of their time thousands of miles away from one another; self-help groups create communities that are at once localized and truly global in their scope. . . . Trust has to be won and actively sustained; and this now ordinarily presumes a process of mutual narrative and emotional disclosure. An "opening out" to the other is the condition of the development of a stable tie—save where traditional patterns are for one reason or another re-imposed, or where emotional dependencies or compulsions exist.

Thus, while a change is in the air, it is important to look into where and how the change is taking place. Is it possible that it is only the individual and his or her interests that drive society? The perspective of Beck certainly pushes us into thinking that such a thing is possible, and the more optimistic Lash asserts that "the motor of social change in this process is individualization." Is this the answer? It is time to give up the sociological project of examining broad social, not individual, structures? Are ICTs just an improvement in the way that we individualize ourselves, or are they (as some suggest) a salvation of the social?

■

The point to be pursued in the following chapters is that, while there is an increasing tilt toward individualism in many contexts, that ritual interaction is still a central function of the collective. Ritual and our sense of the collective are being affected by the use of mediated interaction, specifically mobile communication. While in some situations the use of mediated interaction plays out against the co-present situation, it is also possible to say that mobile communication strengthens the nature of the small group, perhaps even turning it into a clique in some cases.

There is, on one hand, a chance that our situation in society will cause us to drift toward individualism. Indeed, many of the technical solutions available to the individual as well as many ideologies in society also support this tendency. At the same time, there are many mechanisms that are used to bridge differences.

The impulse to reconnect has several levels. There are received forms and there are forms that are minted afresh. For example, language is used—however imperfectly—to connect with others across the gap of individual understanding (Searle and Freemen 1995). In the use of language, we have a great willingness to make sense of others' talk—that is, to be social.

Living in a country where I am not a native speaker, I have almost daily interactions wherein I see greatness of spirit in play. I am, in essence, a continual breaching experiment (Garfinkel 1967). Speaking my pidgin Norwegian, I know that I am mispronouncing words and ignoring important points of grammar at every turn. I have to be willing to try and fail to communicate, and my listener has to be willing to put up with imprecision in the language such as one might expect from a toddler. The interesting thing is that it works. I can order things over the phone, and I can comment on daily life. I can talk a friend through a problem he is having with the software on his mobile phone, and once in a great while I can throw a halfway successful bon mot into a conversation. My ability to do this describes in a small way my ability to speak Norwegian and in a much greater way the largesse of my tortured listeners. In a broader way, however, it comes back to the point that we use mechanisms to hold society together almost in spite of the odds. Despite of the fact that my native Norwegian-speaking interlocutors carry far different linguistic baggage than do I, we are still able to pull off interaction. Further, this means that there are mechanisms for ensuring this. There is the syntax and the grammar of the language that I so imperfectly employ. In addition, there are other types of structures that come into play. There are the expected lines of narration that are often used in interaction. There is the context of the conversation. This means that I can apply a set of expectations within that context in order to help me understand the drift of the interaction. It is OK to talk about job things at work and sports things at a soccer game. My discussing my bank balance with my child's teacher would be surprising; my discussing it with my wife would not be surprising.

■

We may be living in a time when some of the more prominent institutions are on the decline and the pendulum is moving in the direction of the individual. Perhaps the "bricks and mortar" of many institutions are going through troubled times. Institutions such as the church, commerce, and education can wax and wane. In particular, the careers of the larger forms of sociation have a broader impact as they rise and fall. On the way up, they must muscle their way onto the scene; as they wane, they perhaps experience various types of rear-guard actions and power vacuums. As they grow and shrink, there are various types of social turbulence (Bugeja 2005). It is something else to say, however, that the more fundamental institutions of communication, and dare I say ritual, are on their way out. It is possible that they are being reconstituted as a result of the entry of personal ICTs such as the mobile phone. It is possible that another, less physically bound form of sociation is moving onto the scene and is playing itself out in terms of Simmel's "impulse to sociability."

3
Durkheim on Ritual Interaction and Social Cohesion

In the previous chapter I discussed the interaction of technology and social cohesion, an issue about which we clearly worry. This is seen in various discussions of the effects of television, the internet, comic books, and other forms of mediated communication on the ways we relate to one another.

In trying to divine how mobile communication and the internet affect social cohesion, it is important to think about why ritual is such a big deal. It may strike the reader as far afield to raise questions of ritual and Durkheim's study of religion when discussing mediated interaction. While there may be the impulse to see some people's use of mobile communication or the internet as a substitution for religion, that is not the purpose of this discussion. Rather, we are taken to a discussion of ritual—and thereby religion—because of the work of the sociologist Emile Durkheim.

▪

Among graduate students at the University of Colorado in the early 1980s—that is, at the dusk of the structural-functional turn in American sociology—there was the implicit idea that Durkheim was largely a part of that tradition. That is, he was a part of the more conservative line of sociology. The subtext was that Durkheim was to be avoided in favor of either a conflict or an interactionist perspective. From our reading of his work in *The Division of Labor in Society* (1997/1893), he seemed to be legitimizing social stratification. This, of course, was antithetical to the more revolutionary of the graduate students and of little interest to those who chose the interactionist path.[1]

More recently, Durkheim has been reexamined. Randall Collins—often seen as a more critically oriented theorist—began to draw on Durkheim.

As Collins brought to our attention, on one hand Durkheim is the god-father of the structural-functionalist stream in sociology developed at Harvard University by Talcott Parsons.[2] The other strand of Durkheim's work—and the one to be followed in this analysis of small group inter-action via mobile telephone—is the basis for the so-called interactionist perspective. Indeed, in *Four Sociological Traditions* Collins starts his discussion of Durkheim and the Durkheimian tradition with the phrase "We now come to the core tradition of sociology" (1994, p. 181). That is rich praise from a sociologist who has made significant contributions to the critical tradition.[3]

Collins continues by saying the conflict or critical tradition in sociology has provided insight in to the underlying economic and political processes in society by allowing us to see through ideological formations. However, it is the Durkheimian tradition that gives us a lens with which we can see the processes engendered by the symbolic and ritual aspects of society. This focuses us on the central question of sociology—that was also raised in the previous chapter—namely, what holds society together. Bringing this around again to mediated interaction and the mobile phone, we can pose the question as to its ability to support ritual interaction.

Warner and Goffman[4] have drawn on the foundations Durkheim laid. Warner (1957, 1963) applied the notions of tribe and ritual to an industrial city. By doing this, he helped us to take Durkheim out of the background, freshen his insights, and show the relevance of ritual interaction in present-day society. As we will see in the next chapter, Warner's student and research assistant Erving Goffman continued on this path. Goffman moved the focus of analysis from the group to interpersonal analysis. He moved from Durkheim's specialized group ritual to the banalities of everyday life. Goffman's unit of analysis was not the individual or the group; it was the situation.

▪

Durkheim's analysis of ritual as a social glue comes from his work on religion, largely in *The Elementary Forms of Religious Life* (1995). Durkheim developed the idea that religion is not previous to society, but rather the opposite. His insight is that individuals come together to create various social formulations, among them religion. In a restatement of the basic Durkheimian idea, Freeman (1993) noted that society, like religion, is something that we create as individuals in groups. We feel it is true because we believe it to be true. It exists because we believe it to exist. Freeman, incidentally, was sacked from the Church of England for pursuing this sacrilege.

Rituals—and in particular ritual in the sense of mutually focused activity that engenders a common mood in a bounded group—are important because they are the mechanism through which we create and maintain our sense of the collective, in both large and small format. It is here that we come back into the theme at hand. We do the "heavy lifting" of social maintenance through talk, writing, gesture, and (I assert) by talking to one another on mobile phones. The question of course revolves around our ability to maintain social cohesion as we adopt the use of information and communication technologies and thus become less co-present and co-temporal.

The sense of ritual as a source of cohesion arose from Durkheim's analysis of social gravity or social density. He asserted that there is a greater level of interaction within a society as the complexity of the interaction and indeed the complexity of the social structure increases. In society where there is little or intermittent contact the social structure is relatively non-specialized. Individuals can do many different things each carrying out a broad range of essential tasks. As the social density increases—either through the development of urban centers, or through the increase in transportation and communication, there is a specialization in the social structure. Ritual interaction is one of the methods through which the cohesion arises.[5]

For Durkheim, rituals are a way in which the individual is exposed to, and at some level accepts the collective ideas and influences of the broader social order. It is the connection between the laity and the broader drift of the social order—often with a religious hierarchy as the mediators. The function of ritual then is to strengthen and support group solidarity. The point of the ritual is not to embellish the status of the individual, but rather for the individual to celebrate the common status of the group. In this way, the ritual is a commemoration not of an abstract deity, but of the group itself.[6] Rituals affirm the social. The ritual provides a way for us to attach ourselves as individuals to the broader social sphere.

■

It is perhaps odd that Durkheim provides this somewhat revolutionary insight into religion. It is something that we might expect from Marx (with his notion of the opiate of the masses) or Weber (with his ideas on the increasing rationalization of society). Since Durkheim was a son, a grandson, and a great-grandson of rabbis, we might have expected a bit more filial piety (Jones 1986, pp. 12–23). However, in spite of this (or perhaps because of it) he used a considerable part of his career exploring the interaction of religion, ritual, and society.[7]

Between 1902 and the start of World War I, when he was at the Sorbonne, Durkheim wrote extensively on religion and ritual, though he had begun to examine the role of ritual and social cohesion in the mid 1890s (Collins 2005). Starting in 1907 with "La Question religieuse: enquête internationale," it reached its crescendo in 1912 with the publication of *Les Formes élémentaires de la vie religieuse*, known to the English-speaking world as *The Elementary Forms of Religious Life* or more casually as *The Elementary Forms*. Durkheim's work in this area did not go without comment. His position at the Sorbonne meant that his courses were obligatory for many of the students. Thus, it was not without the raising of eyebrows in certain quarters that he forwarded notion that society was responsible for the creation of religion—and not the opposite. This, along with reverberations in the aftermath of the Dreyfus affair, meant that Durkheim's work was noticed, and indeed was the cause of certain turbulence.

It is also worth noting that Durkheim's work on religion was based on second-hand studies of the practices of Australian Aborigines and of Native Americans. Durkheim never traveled widely outside France and never did any field work. Much of the material that is used as a basis for his conclusions was collected by missionaries and anthropologists. The work is based, in a sense, on comparative literature. Durkheim draws on the studies of Boas J. Long, Lewis H. Morgan, Lorimer Fison, and Alfred Howitt. He referred to works by Sir James Frazier (who examined religion in largely the European context) and by Baldwin Spencer and Francis Gillen (authors of *The Northern Tribes of Central Australia*).

Durkheim focused more on the analytical grouping of facts than on the work of gathering them—an approach that was shared by other scholars of the time. Frazier, author of *The Golden Bough*, did not travel much, and Marx spent more time in the British Library than in factories.

Durkheim and Ritual

In the process of examining how mobile communication fits into the firmament of social cohesion and the idea of social unity, the first job is to understand ritual. Thus, we need to take up the interaction between co-present ritual and social structure. How is social cohesion developed and nurtured? Will mediated interaction fray the social fabric, or strengthen it? Although Durkheim did not gather the data firsthand, the notion of cohesion (or solidarity) of society was central to his work. He was concerned with the mechanics various groupings and how they became unified.

Durkheimian ritual is the first step in coming to grips with the issues raised in the previous chapter. In an often-cited passage, Durkheim sees ritual as a way for individuals to come together and form a broader sense of the group:

By themselves, individual consciousnesses are actually closed to one another, and they can communicate only by means of signs in which their inner states come to express themselves. For the communication that is opening up between them to end in a communion—that is, in the fusion of all the individual feelings into a common one—the signs that express those feelings must come together in one single resultant. The appearance of this resultant notifies individuals that they are in unison and brings home to them their moral unity. It is by shouting the same cry, saying the same words, and performing the same action in regard to the same object that they arrive at and experience agreement. (Durkheim 1995, pp. 231–232)

Durkheim begins his argument with the suggestion that religion is a product of social interaction. Religion is not an antecedent to society; rather, it is through social interaction that we develop religion. Religion (read: collective actions of individuals) is one of the processes through which social cohesion is possible. The ritual of religion, with its attendant effervescence, is a basic aspect in the development of cohesion, and it is also embedded in totemic emblems and artifacts.

Durkheim does not see ritual as inflexible and stylized behavior. Ritual is the means through which we engage in a common mood and gain a mutual sense of solidarity. There is a special dynamic here. In a particular setting, various people are individually taken by the spectacle of the situation. It is not enough, however, that they are individually moved. Rather, ritual interaction occurs when the people who are taking part in the situation mutually recognize the common mood of the event. It is by me seeing those near me being moved by the event and reciprocally, they seeing me being moved that we recognize that we are experiencing a common estimation of the situation. This insight is the key to ritual interaction. We have to shout the same cry, say the same words, perform the same action, and engender the same mood. It is through the effervescence of the situation that we realize our moral unity. This is the basis for the cohesive link.

■

As was noted above, religion is not causally previous to society, but rather the opposite. Interaction between groups of persons via the catalytic nature of ritual interaction coalesces into various forms of religion. Thus, it is social interaction that is the basis of social cohesion. It is also the way in which religion is institutionalized. Durkheim writes that "religion is an eminently social thing" and goes on to say that "religious representations are collective representations that express collective realities; rites are ways of acting that are born only in the midst of assembled groups and whose purpose is to evoke, maintain or create certain mental states of those groups" (1995, p. 9). He notes that a society is to its members what a god is to the faithful:

The sacred principle is nothing other than society hypostasized and transfigured; it should be possible to interpret ritual life in secular and social terms. Like ritual life, social life in fact moves in a circle. One the one hand, the individual gets the best part of himself from society—all that gives him a distinctive character and place among other beings, his intellectual and moral culture. . . . On the other hand, however, society exists and lives only in and through individuals. (ibid., p. 351)

At its core, however, cohesion is little more than a belief. Social cohesion is a subtle and intangible thing. It is a shared conviction, and as such it is precarious. It exists in the traditions and in the culture of a group, but it can be unstable. It can be challenged through changes in the material culture. Indeed, the introduction of electronically mediated forms of interaction challenges, in some ways, the received sense of how society ought to function and thus they are the focus of comment.

The interaction between the material and the non-material culture can influence how social cohesion is played out (Latham and Sassen 2004). Thus, social cohesion is not a stable given; it is variable. Durkheim says: "Let the idea of society be extinguished in individual minds, let the beliefs, traditions, and aspirations of the collectivity be felt and shared by the individual no longer, and society will die." (ibid., p. 351) I will revisit this point in the final chapter.

Solidarity is in the minds of the participants. It is through our individual interactions and our willingness to identify ourselves as members of the antelope clan, the "Broncomaniacs," the congregation, the nation or the bowling team that is central. Benedict Anderson's idea of "imagined communities" (1991) and Salen and Zimmerman's idea of the enchanted circle (2004; see also Harvey 2006) play on the same issue. This evokes W. I. Thomas's (1931) dictum that if we believe something to be real, it is real in its consequences. Durkheim says essentially the same thing in slightly more turgid prose: "Their unity arises solely from having the same name and the same emblem, from believing that they have the same relations with the same categories of things, and from practicing the same rites—in other words, from the fact that they commune in the same totemic cult." (1995, p. 169)

Thus, the basic idea is that the belief in collective arises from interaction. But what is the specific mechanism that results in this belief? The answer to this question is, of course, "the ritual."

The Nature of Durkheimian Ritual

Ceremony and ritual provide a context for the development of social cohesion. It is the function of ritual to ensure that a sense of the collective arises from the interaction of individuals. Rituals "set collectivity in motion; groups come together to celebrate them. Thus, their first result is to bring individuals together, multiply the contacts between them, and make those contacts more intimate. . . . And so men feel there is something outside themselves that is reborn, forces that are reanimated, and a life that reawakens." (Durkheim 1995, pp. 352–353)

In an assembled group, there is the opportunity for the individual to give himself or herself over to the collective. It is, however, important to note that Durkheim speaks only of co-present groups and not of mediated groups.

There is no discussion of electronically mediated interaction in Durkheim's works. The words 'telegraph' and 'telephone' are not to be found in his major works. The only mention of newspapers in his major works is a passing reference to their ability to spread the idea of suicide (1951, p. 141). Clearly, mediation in the sense of mobile telephony was far in the future at the time of Durkheim's work. Nonetheless, as we will see below, the insistence on co-presence has been a central issue in the analysis of ritual and social cohesion. It is only in the 1970s that a strain of analysis looking into the telephone arose (de Sola Pool 1971; Wurtzel and Turner 1977).

Durkheimian ritual, in its full sense, provides the individual with a kind of total emersion into another way of thinking and being. It may involve a shift of costume, a special location, the use of music and special foods, unique choreography, and unfamiliar forms of interaction. It often involves the staging of an event by "elders" or other persons who control the apparatus of the event and is staged for a congregation or for new initiates. Most importantly, there is a unity of purpose expressed by companions, and there is the transportation into another "special world" where the mundane and the ordinary are temporarily set aside. Van Gennep (2004/1960; see also Couldry 2003 and Turner 1995) describes this as a liminal process. That is it takes the individual through a threshold from one status to another.

Van Gennep's liminal rituals can, in some cases, be sustained for days and even weeks. It is in these rituals that we see an intensity that is seldom seen in daily life. It is in intense social situation of this kind that, following Durkheim, social cohesion is born and fostered. According to Couldry, this liminality—that is, the crossing of a threshold—is a necessary aspect of ritual interaction (2003).[8]

It is in situations of this kind that Durkheim's "effervescence" arises. The uniqueness of the situation and the high degree of mutual focus cause this effervescence. There may be the common focus on a chant or a dance, or there may be the collective recitation of a liturgy. In this process, the participants see that their colleagues are also active participants in ceremony. There is a common unity that derives from the shared interaction. This process of interaction is achieved when the participants understand their common willingness to show respect for the leaders and the process as well as their common obeisance. The individual's genuflection is, on the one hand, a show of respect to the deity, or its representation. At the same time, it is a sign to others of equal status, that the individual is one of them. He or she is a part of the collective, and so there is the ability forge a sense of their common situation. In short, collective representations arise from the machinations of the ritual (Bellah 2005). There is cohesion. There is sociation.

The power of the ritual can be seen in the use of ritual transgressions that are sometimes tied to initiation rites. The group that gathers in order to initiate new members often uses various forms of hazing and the transgression of norms. The point is that in this setting there is a bond, and perhaps even a distribution of guilt, that serves as the core of the social cohesion. At the same time, these types of ceremonies provide the individual with a whole range of relations (the fellow initiates) and also they define power relations within the group. Thus, they help to structure everyday mundane life. The ritual establishes the relationships and that are again carried over into normal life.

Rituals also help us to distinguish between the in-group and the out-group. (See Duncan 1970, p. 208.) Ritual results in a "fusion of sentiments." Through ritual interaction, individuals realize their common "moral unity." The individuals are fused into a unity that is different from and perhaps opposed to those who are not included in the process. Indeed this is an important point. The ritual defines a boundary between those who are members and those who are not. With the liminal rituals of van Gennep, there are those who have been initiated and those who have not. Even in less explicit ritual interaction such as those examined below ritual can be used as a kind of moral boundary.

▪

The specific content of the ritual is not necessarily relevant. In the previous chapter, I discussed the way in which ritual can be seen as a kind of obsessive and repetitive behavior that always has the same features. While that sense of the word 'ritual' may have currency and meaning in some situations, Durkheim used the word differently. Indeed, he felt there was a plasticity regarding the concrete forms used in the ritual. The point is not the steps and stopping points on the Via Dolorosa, the particular preparation of the special meal, or the way the fans cheer when team enters the stadium. Rather, it is how mutually understood interaction affects cohesion:

. . . several rites can be used interchangeably to being about the same end. . . . This interchangeability of rites demonstrates once again—just as their plasticity demonstrates—the extreme generality of the useful influence they exercise. What matters most is that individuals are assembled and that the feelings in common are expressed through actions in common. However, as to the specific nature of these feelings and actions, that is a relatively secondary and contingent matter. To become conscious of itself the group need not perform some acts and not others. Although it must commune in the same thought and the same action, the visible forms in which this communion occurs hardly matter. (Durkheim 1995, p. 390)

We should not necessarily focus on the nature of the liturgy used by a group. What is important is that the group becomes aware of itself and, in doing so, develops a sense of unity.

The Totem as the Reservoir of Group Solidarity

With the Durkheimian ritual, there is the need to understand how we might sustain the sense of the group. It is here that the role of the totem comes into play:

Without symbols, moreover, social feelings could have only an unstable existence. Those feelings are very strong so long as men are assembled, mutually influencing one another, but when the gathering is over, they survive only in the form of memories that gradually dim and fade away. Since the group is no longer present and active, the individual temperaments quickly take over again. Wild passions that could unleash themselves in the midst of a crowd cool and die down once the crowd has dispersed, and individuals wonder with amazement how they could let themselves be carried so far out of character. But if the movements by which these feelings have been expressed eventually become inscribed on things that are durable, then they too become durable. These things keep bringing the feelings to individual minds and keep them perpetually aroused, just as would happen if the cause that first called them forth was still acting. (Durkheim 1995, pp. 232–233)

The totem is an object or icon into which the energy of the collective ritual is symbolically invested. That is, the mutual awareness—and the resulting effervescence—experienced by the participant's is symbolically attached to emblems and symbols. The totem can be an image of the clan's totemic animal, flower, a symbol (such as a cross), or an image of the moon or a star. It is a point of identification and a central defining element in their ritual social interactions. There is also the ability to extend the symbols to other realms that include the spoken word and the gesture.

The totem becomes the symbolic reservoir for the group's social interaction. It gains a sacred role. In the absence of the social interaction, the totem serves to remind the individuals of their position with respect to the other members of the group. Obeisance to the symbol becomes a way to identify with the group. The totem can also mark boundaries between groups. With bear-related dress, jewelry, or body art, for example, members of a clan or a group can show allegiance to the bear. Regardless of its form, when the individual is not engaged in the direct ritual, the totem becomes the substitute and so respect for the totem is the way to show allegiance.

As with any reservoir, however, there is leakage. The totem eventually loses its ability to remind the members of the fervor associated with the original situation where in it was endowed with its symbolic power. Thus, it needs to be periodically rejuvenated. Rejuvenation is accomplished by the reunion of the group and the performance of rites or a liturgy that includes the totem. The periodic meeting of the group, be it a political party, religious domination or economic group, is needed to refresh and renew their common bonds and to rejuvenate the totem. The rejuvenation of the totem—and hence the group—is accomplished through various forms of performance. There can be the eating of the totemic animal, participating in the proscribed singing or dancing, the initiation of new members, or perhaps by meeting to share stories of how the cross-town bullies were recently humiliated.

As we will see below, the fate of the totem is one of the ways to distinguish Durkheimian ritual from that of Goffman and Collins. I assert that the totem has been replaced by perpetual contact, to borrow an apt phrase from James Katz and Mark Aakhus (2002a). In the Durkheimian system there was the question of how to perpetuate the sense of group solidarity across time between ritual events. Goffman felt that ritual interaction was so enmeshed into everyday life that that there is quite literally no mention of totems in the work of Goffman.[9] Bringing this discussion back to the issue of mobile telephony I will posit that mobile communication further extends the possibilities for interaction beyond the co-present. Thus, while co-present interaction is that realm where solidarity is quite often founded, the glow of the event and the glow of solidarity within the group can be, and indeed is in many cases kept alive via mobile communication. In this way, mobile communication obviates totems.

The Agency of Society

Up to this point, I have been discussing the way in which the individual becomes attached to the group and how through the machinations of the ritual they are held there. Durkheim discusses the actual agency of society: "Precisely because society has its own specific nature that is different from our nature as individuals, it pursues ends that are also specifically its own; but because it can achieve those ends only by working through us, it categorically demands our cooperation." (Durkheim 1995, p. 209)

This notion is an interesting turn, since it says that society and individuals are, in some senses, separate entities. The social agency issue tells us about the backdrop. The massively concrete nature of society—although developed through the interactions of previous individuals—is that milieu in which current social activities take place. In this way, it prefigures the interaction and it sets the expectations and the parameters. Collins (1998, p. 47; see also Deacon 1997) talks about language itself a being the product of ritual, and Anderson (1991) discusses "unisonance" in the singing of national anthems.

Society is prior to individuals and has a major hand in shaping them. Foreshadowing the discussion in the following chapters, individuals carry over what Collins calls emotional energy from one situation to another and through this mechanism form the core of social interaction.

As we move into and out of different situations, we look to the broader sense of the context in order to glean clues as to what is appropriate in a given situation. The massiveness of this is largely received, but our interaction in the in the social order embeds us in the broader sweep of things.

On the one hand, there are people as individuals; on the other hand, there is something called society that is made up of those same individuals but in their collective sense and with the encrustations of previous individuals impinging on their actions:

Society requires us to make ourselves its servants, forgetful of our own interests, and it subjects us to all sorts of restraints, privations, and sacrifices without which social life would be impossible. And so, at every instant, we must submit to the rules of action and thought that we have neither made nor wanted and that sometimes are contrary to our inclinations and our most basic instincts. (Durkheim 1995, p. 209)

This sounds rather repressive. If we were to substitute 'ideology', 'despot', or even 'religion' for 'society' in the preceding quotation, it would be a call to arms. It might motivate those who are inclined to such to, for example, join the military in order to free themselves and society from this oppressive shadow that hangs over the landscape. However, there is no such call since the social enjoys legitimacy that is born of our collective willingness to participate in it. This agency is not necessarily repressive. Since there is, in general terms, legitimacy for the social order, we willingly participate in it, and through that participation we are able to carry out our individual purposes, but we vicariously support the social order.

This may bring to mind the metaphor of ants, which when foraging for food leave behind a scent trail that others may follow. Some ants establish trails; many follow them. Those that come later do not, however, simply tread the same path. Rather, as they pass, they also leave behind their scent, and thus they contribute to the trail qua institution. In collectively going about our daily activities, we also participate in and further develop a sense of the social. Because of the mutual nature of society, it has a hold on our consciousness that individual states of awareness cannot match:

> Because these ways of acting have been worked out in common, the intensity with which they are thought in each individual mind finds resonance in all others and vice versa. The representations that translate them within each of us thereby gain an intensity that merely private states of consciousness can in no way match. Those ways of acting gather strength from the countless individual representations that have served to form them. It is society that speaks through the mouths of those who affirm them in our presence; it is society that we hear when we hear them; and the voice of all itself has a tone that an individual voice can not have. (Durkheim 1995, p. 210)

Thus, there is the sense of society as having a form of agency. This agency describes the general context in which the individual carries on their daily activities. It also tells us why, when there are new wrinkles in the physical culture (such as the introduction of mobile phones), there is turbulence. On one hand, the new way of doing things makes it necessary to re-examine the broader agreement as to how interaction should take place.

There are elements of the social order that are received and that we must learn to revere. At the same time, we are, as individuals, also co-authors of the collective order—but not necessarily in the sense that we each individually develop new forms of social intercourse or found new social institutions. Rather, we participate in the evocation, in the maintenance, and on very rare occasions in the creation of social forms.

Of course, this is not to deny that there is individual agency. Though there may well be a drift toward individualism, the social order still has a major role. The basis for the continued willingness to defer to the social is, according to Durkheim, that it is an object of respect—not only respect for an abstract notion, but respect for a phenomenon for which we are mutually responsible. We have received the encrustations of previous generations, and we have contributed to the ongoing process of society. We have received such broad social institutions as language, forms of interaction, and monetary systems. We also have received specific senses of courtesy, notions of how friendships are formed and maintained, and such recently coined things as slang-based greeting sequences and unique phrases used in text messages.

From Macro to Micro, from Co-Present to Mediated

This chapter has traced Durkheim's notion of the ritual as the wellspring of social cohesion. Durkheim examined how ritual contributes to a social structure that becomes massive and prefigures further social interaction. He studied this in the context of aboriginal religion.[10]

The notion of ritual is, in Durkheim, tilted toward the special event, the celebration, the cyclical rite. However, since this book is about rituals that often take place at a different level of magnitude, we must draw a connection between the more segregated sense of 'ritual' and the more everyday sense used by later observers. Interestingly, Durkheim leaves the door open for Goffman—and for me—when he states that it is not only in the formal times of interaction that ritual comes into play:

There is virtually no instant of our lives in which a certain rush of energy fails to come to us from outside ourselves. In all kinds of acts that express the understanding, esteem, and affection of his neighbor, there is a lift that the man who does his duty feels, usually without being aware of it. But that uplifts and sustains the feeling he has for himself. Because he is in moral harmony with his neighbor he gains new confidence, courage, and boldness in action—quite like the man of faith who believes he feels the eyes of his god turned benevolently toward him. (1995, p. 213)

Here Durkheim at least flirts with the idea that mundane life is also infused with minor rituals. The study of social interaction can turn on examination of how completely the rituals and totems command the common attention and how completely they entrain individuals. There is a hint that it is not only the broad social rituals that result in totemic symbols. Durkheim sees that it is possible to generalize this process: "In all its aspects and at every moment of its history, social life is only possible thanks to a vast symbolism. The physical emblems and figurative representations with which I have been especially concerned in the present study are one form of it, but there are a good many others." (ibid., p. 233)

With these thoughts, Durkheim seems to move away from the idea that ritual (or at least the effects of ritual) is experienced only in segregated situations. He seems to be saying that ritual is apparent at many points in our daily lives. As in the work of Goffman, this is seen in the ways we greet, interact, and part. It is seen in everyday events as mundane as paying a toll at a toll booth and navigating through a grocery store with a shopping cart.

▪

In chapter 8, I will outline how these mechanisms are also seen in mediated forms of interaction. A rich and growing tradition suggests that mediated interaction will influence the need for co-presence as outlined by Durkheim. Indeed, in some circles—for example, among internet-based gamers—it is somehow seen as stodgy and old-fashioned to think of co-presence as the main motor of community. Internet gaming, instant messaging, web-based groups, social networking sites such as Facebook, and weblogs all represent new ways of socializing (Harvey 2006). While some of this assertion is based on a false sense of technology's place in society, it is also clear that sociation as described by Durkheim can be conducted in these contexts.[11]

But these conjectures raise questions. For example, what is the role of mobile communication in maintaining social cohesion between co-present events? Can we use it to build up to and to tail off from "flesh meetings"? Ito (2005b) describes this as lightweight interaction, and Licoppe (2004) looks at the remote interactions as a kind of continual contact. This may suggest that the idea of the totem is being changed. Howard Rheingold has referred to the mobile phone as a modern talisman. It is an odd variation on the Durkheimian totem. It is not simply the repository of the symbolic energy from our last meeting; it is also an oracle through which we can communicate with the social group, and it is also a symbolically important physical artifact.

It is important to remember that Durkheim had, in a limited way, taken the notion of ritual into the mundane machinations of everyday life. In addition, mediation was nothing to Durkheim. Thus, we are left his insight into the collective production of social rituals. In *The Elementary Forms* the reader is encouraged to imagine only the sweaty co-presence of a ritual. Earlier I quoted a passage in which Durkheim paints a picture of individuals in a ceremony "shouting the same cry, saying the same words, and performing the same action." There is undeniably the construction and development of social cohesion here. It is difficult, however, to imagine this kind of social cohesion being developed through mediated interaction.

Is it possible to engender social cohesion in a normal mundane interaction? Does the situation have to be co-present? Is there the potential for everyday ritual and is there the opening for mediated ritual? The answer to the first question seems to be a definite Yes. Indeed, Goffman, Collins, and a raft of others outline this possibility. The second question seems more tenuous. Present-day sociological theory is based largely on the idea of co-presence. With the infusion of mediated interaction in society, will the center hold? Will the social density that mediated interaction provides allow for the development of social cohesion? ICTs probably will result in new forms of anomie. It can be imagined that they will also provide new channels for making and cultivating social contact.

4

Goffman on Ritual Interaction in Everyday Life

In the previous chapter, I outlined Durkheim's ritual interaction and its relationship to cohesion. Ritual interaction takes place when the participants in a situation share a common mood and recognize their mutual entrainment. The actual effervescence of the situation is the catalyst for the development of cohesion. These interactions develop into a broader sense of solidarity where the individuals share the same perspective, talk about the same issues, and submit to the same ideals. According to Durkheim, in order to be merged into a single cohesive whole individuals must shout the same cry, say the same words, and perform the same action before they "arrive at and experience agreement" (1995, pp. 231–232). That is, they must experience a collective co-present ecstatic event that is highly focused and that has, in the words of Collins, a high degree of entrainment. In short, Durkheim provides us with the sense of how ritual interaction can serve as a kind of social glue.

Although Durkheim helps us understand how ritual results in social solidarity, his analysis is, in many respects, too broad for understanding the role of everyday ritual, such as ritual associated with mobile telephony. Where Durkheim looks at the periodical large-scale social rite as a way to generate social cohesion, mobile communication is a small-scale mundane interaction that is often used for various types of functional coordination as well as phatic interaction.

In the case of mobile communication, rather than examining a ritual peopled by a large number of celebrants, we are faced with examining two people talking on their mobile phones. Rather than uttering the same cry and performing the same action in co-present unison, we are faced with people who are not even within shouting distance. In order to turn Durkheim's insight into a tool for the analysis of telephonic interaction, we must scale it down.

In the writings of Erving Goffman we find insightful observation of social interaction. More important, we find an astute analysis of social ritual at the interpersonal level. Following in the Durkheimian tradition, Goffman reapplies the idea of social ritual to everyday activities. He shows us how to approach the humdrum activities of co-present daily life. He shows us how they influence social solidity, and he shows us how to appreciate in-group interactions, common mood, and entrainment on a small scale. Unfortunately, Goffman was only marginally interested in meditated interaction. Though he discusses the telephone at various points, he only hints at its role in social cohesion.

Goffman's Links to Durkheim

As I mentioned in the previous chapter, Durkheim examined the role of religion in the lives of Aborigines. Moving from early-twentieth-century France to the mid-twentieth-century United States, let us pick up this thread. Goffman, following in the footsteps of his teacher W. Lloyd Warner, re-scaled Durkheim's insights. Whereas Durkheim focused on ritual as related to the exercise of religion, Warner and Goffman broke free of the religious dimension. That enabled them to see the role of ritual in mundane everyday life. Whereas Durkheim saw ritual as a periodic practice, Goffman saw it in daily interaction. Whereas Durkheim saw rituals arranged by priests and leaders for the benefit and instruction of the initiates and followers, Goffman often saw individuals as both authors of and participants in their ritual interactions.[1]

Warner was a particularly fitting link between Durkheim and Goffman. In echoes of Durkheim, his early work was an anthropological examination of Australian Aborigines (1957). Warner, like Durkheim, examined the functioning of solidarity of this culture. Unlike Durkheim, however, Warner studied the culture firsthand, doing field work in Australia. Warner then moved to apply the same techniques of study in his own culture in the "Yankee City" studies (1963). In this way, he made the first steps in moving from the study of cultures as remote abstract entities, as was the case in Durkheim, to studying his own local culture. Goffman followed the lead of Warner, and indeed he was Warner's student at the University of Chicago in the late 1940s. As we will see below, he took the step of scaling down the analysis to small-group interaction.[2]

In addition, based on the insights of his intellectual mentors, Goffman looked at the role of ritual in social cohesion.

There is a relatively uncluttered lineage between Goffman and Durkheim.[3] It is easy to see the Durkheimian interest in social cohesion and the Warnerian methodological techniques. Indeed, Goffman was explicit in his homage to Durkheim. In the conclusion to the essay "On deference and demeanor" he points to the connection between his work and Durkheim's. He then carries the analysis of Durkheimian religion to the level of the individual and their behavior and movements in everyday life:

> . . . I have suggested that Durkheimian notions about primitive religion can be translated into concepts of deference and demeanor, and that these concepts help us to grasp some aspects of urban secular living. The implication is that in one sense this secular world is not as irreligious as we think. Many gods have been done away with, but the individual himself stubbornly remains as a deity of considerable importance. He walks with some dignity and is the recipient of many little offerings. He is jealous of the worship due him, yet, approached in the right spirit, he is ready to forgive those who may have offended him. Because of their status relative to his, some persons will find him contaminating while others will find they contaminate him, in either case finding that they must treat him with ritual care. Perhaps the individual is so viable a god because he can actually understand the ceremonial significance of the way he is treated, and quite on his own can respond dramatically to what is proffered him. In contacts between such deities there is no need for middlemen; each of these gods is able to serve as his own priest. (Goffman 1967, p. 95)

Rituals are a source of cohesion and are carried out in small-group interaction. Of particular interest with regard to mediated interaction, such as with the mobile telephone, there are some hints in Goffman that ritual interactions can be mediated. While firmly based in the analysis of physically co-present situations, Goffman enticingly suggests that rituals can carried out if there is the "perception" of co-presence.

The Situation as the Unit of Analysis

Much social research focuses on the individual or (as with social network analysis) on the group. Whereas others examined the broader social landscape, Goffman's unit of analysis was the situation (1963a, p. 197). It is by examining the way that people greet and part, how they comport themselves at restaurants and on the street—and dare I say interact on the telephone—that we see people working out their social interaction.

It is in bounded interaction among a small number of people that we see the social machinations of society.

Goffman had the ability to see how individuals enter into, carry on, and eventually extract themselves from situations. It is in the situation that we can see the use of Goffmanian ritual. It is in the local situation that we use various ritual devices from our storehouse of social effects in order to play our part. We draw on well-used strategies when proffering a greeting. We utter yes-I-am-paying-attention sounds such as "yeah," "umhum," and "oh" at judicious points in the conversation, and we laugh at the jokes or show appropriate signs of sorrow when unfortunate events are discussed. We also draw on different stratagems when extracting ourselves from the social interaction. We are the ones doing the stage management, as there is no priest or leader directing the action.

We can carry out a "chat" in which there is light repartee between two or more individuals bounded by greetings and parting rituals. In addition, the pace of the interaction can be guided by the use of social devices such as "What did you do last weekend?" and "What about this weather?" Indeed, the interaction and flow of a chat can be guided by the management of the briefest of all rituals: the glance (Goffman 1963a, p. 101). By observing how even a fleeting look is accorded meaning and is indeed a kind of focused social interaction, Goffman shows how far from Durkheim's rituals he is willing to go. Indeed, the glance of an authority figure may be enough to bring us back into line when we are up to some mischief, and the glance of a lover may be enough to start some mischief. By examining these social devices, Goffman challenges us to consider the minimal sense of entrainment and the minimal sense of a ritual. In Durkheim, there is elaborate staging of rites. In Goffman, we see the extreme of micro interaction: "The gestures which we sometimes call empty are perhaps in fact the fullest of all." (1967, p. 91) It is in these gestures that we construct and maintain the social order. According to Goffman (ibid.), these are "the bindings of society."

As Durkheim suggested, the central characteristic is the ritual involvement of the participants, however fleeting and however imagined.

Interpersonal Ritual as the Locus of Social Interaction

Though there are intellectual links between Goffman and Durkheim, Goffman eventually developed his own approach to studying social interaction and his own insight into the mechanisms of social situations. The most important point of distinction is that it is not necessarily in religious or large-scale gatherings that ritual cohesion is built up. Rather, it is in everyday interpersonal interaction. For Goffman, ritual is not a separate occasion, nor is there the need for totems that bear the symbolic cohesion of the group. Since ritual is an ongoing part of everyday life, there is no need for separate physical artifacts. This is not to say that particular behaviors do not have symbolic meaning. They are, to a greater or a lesser degree, the content of daily social interaction. Shaking hands, waving goodbye, and opening doors for others all have both functional and symbolic meanings.

Mundane interpersonal rituals are to be seen in all phases of life, and in a largely secular society these rituals bear the weight of engendering social cohesion:

> In contemporary society rituals performed to stand-ins for supernatural entities are everywhere in decay, as are extensive ceremonial agendas involving long strings of obligatory rites. What remains are brief rituals one individual performs for and to another, attesting to civility and good will on the performer's part and the recipient's possession of a small patrimony of sacredness. What remains, in brief, are interpersonal rituals. (Goffman 1971, p. 61)

Here Goffman explicitly turns his attention away from the Durkheimian focus on religion and the large gathering as the locus of ritual interaction. Occasional ceremonies that include extended liturgies are replaced by continual interactions between individuals. Ritual is not to be found only in observations of the winter solstice, in the Mass, or at weddings. Ritual is also to be found whenever there is even the most momentary sharing of mood and the recognition of mutual engagement in a situation or activity.

If a society is to be maintained, it must socialize its members to be "self regulating participants in social encounters" (Goffman 1967, p. 44). It is through the use of ritual that this is to take place, and it is through this attachment that the individual feels that he or she is a part of the whole. It is through this process that individuals develop their sense of pride, honor, and poise. As I noted in the previous chapter, Durkheim showed some openness to this line of thought. He wrote that there is "virtually no instant of our lives" in which we are not cognizant of the ritual interaction with others (1995, p. 213). But Durkheim did not develop this line of thought. Goffman later suggested that Durkheim had looked at the forest, whereas he (Goffman) was looking at the trees.

In the passage mentioned above, we see that Goffman saw the individual as having a special value in social interaction. He notes that we perform the rituals "to and for another." Thus, it is the individual and their responsibility to the situation that has taken on a central role in the symbolic panoply. He notes that as other "gods have been done away with" the individual has remained. This is of interest when we consider interaction with technology. One line of though is represented by Meyrowitz's (1985) analysis of the social dimension of television. Meyrowitz looks at the ways in which television (both the device itself and the actors it brings into our homes) has a social presence. Selberg (1993) takes this a bit further by examining the rituals of watching the news in Norway, where until the early 1980s only one television channel was available. The national news, broadcast at 7 P.M., was in essence a national event shared in the temporal but not the geographical sense by many Norwegians. Selberg describes how this became a metric for determining the pace of various tasks in the home. It was late enough in the evening so that the evening meal was over and perhaps the dishes were washed. The family could then brew some coffee and perhaps have some cookies prepared, placed on a tray, and carried into the living room. The family would then gather around the television, serve coffee, and prepare to watch the presentation of the day's events. In this way the TV, and the particular news program, provided a kind of ontological security. In spite of the fact that the most horrible murders, wars, and pestilence were being presented, the fact that the family was safely gathered together and sharing the interlude provided a sense of security and continuity.

Selberg's analysis follows in the best Goffmanian/Durkheimian tradition. All the elements are in place. The preparations are made, the group is collected, the context for interaction is defined and carried out, and a sense of social solidarity arises from the interaction. The interesting point here is that an otherwise innate object, the television, has a significant animating force in the whole tableau. Indeed the television is a quasi-social actor when seen in this way. With the growth in the number of TV channels in Norway, this viewing pattern has lost much of its salience. Indeed this development is often drawn on as a metaphor for the increasing fractionalization of society.

The other line of theoretical development here is that associated with the so-called domestication approach. Again, this approach to understanding the adoption and use of various artifacts plays on some of the same themes described here (Silverstone et al. 1992; Haddon 2001). Various objects, such as mobile phones, come into the consciousness of the individual, are evaluated, are purchased, and come into use. In this process, individuals are in a more or less continuous process of trying to place the object into the context of their everyday lives. Where is it appropriate to use it, how it should be displayed, and when it should and should not be used are all issues of deliberation. As others in our social circle find out that we too are consumers of a particular artifact, their estimation of us changes. Their perceptions of the object and their perceptions of our display and use of the object, whatever it might be, become parts of their understanding of who we are. These insights affect their definition of us and influence the unfolding of the interaction. In the rubric being developed here, the artifact in some ways forms the interaction.

■

Goffman frames the nature of the individual in social situations by examining the concepts of deference and demeanor (1967, pp. 57–58). On the one hand, we command deference[4] in our interaction with others. At the same time, we use various forms of demeanor or comportment—that is, allowances in our estimation and treatment of others—in our presentations of self.

Deference and demeanor are particularly easy to observe when we are making salutations, proffering complements, making invitations, and providing minor services (ibid., pp. 72–73). They are seen in the performance of etiquette. They are seen in our acceptance of virtual back-stage areas wherein, for example, we allow others the space in which to rearrange their facade by turning a blind eye when needed. It is also possible to examine deference and demeanor when thinking of availability signs, adjustments, and the provision of status information. Out of consideration for others, we attend to our back-stage activities before coming onto the scene and we ask that others cast a blind eye toward the flaws in our presentation. Indeed, Collins (2004a, p. 24) notes that Goffman's discussion of deference and demeanor is the foundation upon which the more cited analysis of front stage and back stage is based.

It is when deference and demeanor are threatened that we see the consequences of their existence. It is when the back-stage activities threaten to "enter stage left" that we need to recompose our self. It is in these situations that we ask, and in most cases receive, the forbearance of audience in the reestablishment of the facade.

Breaches in etiquette provide insight into the nature of social order. Goffman wrote extensively on persons who for one reason or another are unable to fulfill the expectations of social interaction in his book *Asylums* (1961). This book can be read not as a discussion of the individual's challenges when confronted with social interaction in a "total institution," but rather as a discussion of the social process of carrying off an adequate social presentation of self when faced with actors who can only partially fulfill the expectations of their role. Patients' refusal to observe the usual social cues, or their use of common social devices in inappropriate situations, gave Goffman the insight as to how important these were in broader social interaction (1967, p. 62).[5]

Goffman also examined the marking of in-groups and out-groups (1959, p. 162; 1967, pp. 34, 75). In small-group interaction there is a need to define those who are accredited and those who are outside the normal sphere of interaction. As Duncan (1970, p. 208) notes, ritual interaction facilitates the definition of in-groups and out-groups.

Breaches, and our sometimes hurried stratagems for covering them, expose the sense that the social order is somehow in danger. Indeed Goffman's analysis of embarrassment turns on this issue (Goffman 1967, p. 97). When we are somehow placed in a situation wherein our own sense of our facade and its actual state part from each other, we are embarrassed until we are able to deploy some strategy to restore the two divergent story lines.

Just as the residents of Goffman's St. Elizabeth's hospital generated novel social situations, the mobile phone can ring, buzz, or burble with a surprising range at the most awkward moments. In this way, it breaks into the existing flux of the situation and presents us with more or less involving side engagements. This aspect of mobile communication has resulted in a minor industry of relating ever more interesting examples of where a mobile phone has been used. There is also a line of academic analysis of the issue (Ling 1997; Love 2001; Ling 2004b; Monk et al. 2004). We are in the process of developing ways to shield these interruptions. Receiving a text message, for example, is less intrusive than a receiving a call, and moving into a less-traveled section of the public space is less disturbing than carrying out a call in the middle of the scene. We are, in short, only starting to develop ways to block the idea that we are not properly engaged in a particular situation.

When, for example, a mobile phone rings in the middle of a funeral, we see the deployment of strategies for dealing with deference and demeanor. Obviously the mobile telephone presents a particularly rich set of opportunities to study this kind of work in action (Ling 1997).

The use of the device means that we are presenting ourselves on two front stages simultaneously and we are left to juggle between the sensibilities of our two audiences. The staging of a telephone call complicates this issue, since there is potentially more than one public. This has been referred to as a double front stage (ibid.). Who is included in the conversation and who is excluded from it is, in most cases, obvious. The telephone itself defines the interlocutors. In certain stagings, however, someone may interject himself or herself into the conversation (Lohan 1996), or one may become an unwitting observer of a telephonic interaction (Ling 1997), or one may wrongly think that one is part of a conversation circle (Love 2005).

Totems

Goffman's interpretation of Durkheim does not carry with it the notion of the totem. Indeed, there is no mention of the word in his oeuvre. This does not mean, however, that Goffman does not discuss investing artifacts with symbolic meaning. In *Asylums*, for example, Goffman describes how inmates use an "identity kit" for the management of their personal facades (Goffman 1961, p. 20). Such a kit might include soap, needles for sewing and repairing clothes, and combs. The identity kit can be seen as a kind of personalized and very restricted totem. It has a functional side, but it also is an assertion of personality in an otherwise "total institution." It is a set of artifacts with which the individual constructs a sense of self. In a "total institution," the individual suffers the mortification of knowing that society does not trust him or her with simple everyday items and a place where they can be kept. In this way, a comb or a bar of soap gains status as a symbol of identity.[6]

This is a rather serious re-rigging of Durkheim's notion of the totem. The energizing of the object or the mascot in a common rite is not there. The collective reverence and awe afforded by viewing the object is not to be found. Still, to reapply the ideas of Durkheim to Goffman is to recalibrate and scale down the insights from the macro to the micro.

The issue of investing objects with symbolic values is also seen in Goffman's notion of the individual. Rather than thinking of the totem of the bear clan or the sunrise cult, Goffman says that it is the individual that is symbolically central. It is our interpersonal interactions, our use of deference and demeanor that fill the role of the totem as seen, for example in the use of gesture or forms of courtesy. The ritual of shaking hands becomes the reminder of our social position. Instead of relying on a symbol or an object showing how "they commune in the same totemic cult" (Durkheim 1995, p. 169), it is done through various gestures and utterances of common civility. This is not to deny that physical objects and their representations do not have a role in the ordering of mutual expectations. Indeed physical objects can be the locus of reverence.[7] These are, however, more watered down that we might expect from reading Durkheim.

Another point of contrast is that in Durkheim the ritual interaction is used to restore its symbolic status to a totem.[8] In Goffman, however, the ritual interactions are so thoroughly enmeshed in everyday activities that the role of the totem as a central repository of these emotions may be passé. The continual need to be social, to greet, to interact, to converse, and to carry out parting rituals obviates the need for the totemically based ritual. We are, in effect, continually recharging the symbolic value of our social links. It is tempting to read the use of text messaging in this light. This form of interaction is quite often shorn of the usual "niceties" of written interaction. Salutations and closings are not necessary. They often do not contain the small talk associated with a telephone call or the introductory part of a meeting. They do, however, have a phatic content. They are random positive feedback. In addition to whatever instrumental content, they carry out the social task of integrating the sender and receiver (Ling 2005c). These small rituals are so common and so omnipresent for some people that they replace, or at least drown out, the functions of the more traditional totem. While the broader rites that do remain and give a broad direction to our sense of status and identity, i.e. our sense of national or religious identity, it is the daily and all-encompassing Goffmanian ritual interactions that perpetually remind us of our position in society.

The mobile phone is a part of this and, in an interesting way, it extends the ritual reach of society in that it extends the access of our nearest friends and family. In that Goffman talks about daily life as infused with ritual interaction, the mobile phone extends the possibilities for this kind of interaction, particularly among friends and family. Thus, where Goffman was more agnostic as to who the recipients of the small courtesies might be, mobile communication means that we are in perpetual contact with our "clan." Thus, we do not only have to go through the gestures of interaction with strangers or those with whom we have an attenuated relationship. In addition, the device extends the reach of parents, children, and friends. Rather than relying on a totemic representation of a social circle, they are, in the terminology of Katz and Aakhus (2002a), in perpetual contact.

The Staging of the Telephone Call

Goffman is a fruitful source of insight into the use of mobile telephones in co-present situations. The question remains, however, as to how his very physically co-present analysis can be applied to mediated situations.

By and large, Goffman is a "face-to-face" man. He says that the full notion of a situation includes "the sense that they [the interactants] are close enough to be perceived in whatever they are doing, including their experiencing of others, and close enough to be perceived in their sensing of being perceived" (1963a, p. 17; see also 1967, p. 1). Indeed, the subtitle to his book *Interaction Ritual* is *Essays on Face-To-Face Behavior*. Later in that book he writes:

The rules of conduct that bind the actor and the recipient together are the bindings of society. But many of the acts which are guided by these rules occur infrequently or take a long time for their consummation. . . . Wherever the activity and however profanely instrumental, it can afford many opportunities for minor ceremonies *as long as other persons are present*. Through these observances, guided by ceremonial obligations and expectations, a constant flow of indulgences is spread through society, with others who are *present* constantly reminding the individual that he must keep himself together as a well demeaned person and affirm the sacred quality of others. The gestures which we sometimes call empty are perhaps in fact the things of all. (1967, p. 91, emphasis added)

This statement also underscores Goffman's reliance on physical co-presence. His orientation to co-present situations can also be seen in his discussion of "social establishments." In the concluding chapter of *The Presentation of Self in Everyday Life* he does a good job of laying out the topographical nature of small-group interaction. In essence, he characterizes physical co-presence as a requisite element of a situation:

> The social establishment is any place surrounded by fixed barriers to perception in which a particular kind of activity regularly takes place. I have suggested that any social establishment may be studied profitably from the point of view of impression management. Within the walls of a social establishment we find a team of performers who cooperate to present an audience a given definition of the situation. This will include the concepts of own team and of audience and assumptions concerning the ethos that is to be maintained by the rules of politeness and decorum. We often find a division into back region, where the performance of routine is prepared, and the front region where the performance is presented. Access to these regions is controlled in order to prevent the audience from seeing the back stage and to prevent outsiders coming into a performance that is not addressed to them. (Goffman 1959, p. 238)

There is little hint that mediated interaction is in Goffman's thoughts. To some degree, this is a legacy of his times. Although he worked in a period that included telephony, it was definitely not the form of mobile interaction that we experience today. This perspective and the content of his writing certainly seems to limit the expectation that Goffman can be reapplied to telephonic interaction. There is, however, a small glimmer of hope in the previous citation.

Although he certainly did not intend it so, Goffman's suggestion that the establishment is surrounded to "fixed barriers to perception" can, if read with a certain willingness, be broadened to include telephonic interaction. I am interested in exploiting this chink in the discussion. I am interested in thinking about how the "barriers to perception" do not necessarily limit interaction to the physically co-present but only limit it to the perceptually co-present. I am interested in Giddens's suggestion (1986, p. 68) that there are dimensions of Goffmanian co-presence that can be mediated by, for example, the telephone.

Mediated interaction makes cameo appearances at certain points in Goffman's work. Goffman writes, for example, about other forms of ceremonial interaction, such as "gifts, greeting cards and salutatory telegrams and telephone calls," as a medium used to affirm relationships that are not physically co-present (1963a, p. 102, n. 38). The regularity of these interactions is a function of the group. Spouses on business trips were seen to need a higher frequency of such contact than, for example, those who are "on the Christmas card list."

It is interesting to note that telephonic interaction is mentioned in the signature Goffmanian concepts of front stage and back stage. The telephone is explicitly a part of the back stage: "Here [in the back-stage area] devices such as the telephone are sequestered so that they can be used "privately." (Goffman 1959, p. 112) Goffman also speaks of various ruses that might be used in order to cover over telephone conversations in public areas (ibid., p. 186). In his later work *Behavior in Public Places*, Goffman seems to bring the telephone onto the quasi-front stage when he observes that when one member of a co-present conversation pair accepts a phone call the other member often takes the pose of civil inattention. When it is a three-part co-present conversation and one of the three participants receives a call, the remaining two often lapse into a "limp" conversation (Goffman 1963a, p. 158). But is clear that Goffman is talking about telephony in a different era. There is, for example, talk of the now-rare dormitory telephone, and of the fact that lower-status residents of the dormitory often had responsibility for answering the phone and higher-status residents allowed themselves to be called to the phone several times in order to broadcast their popularity.

Another enticing reference to telephony comes when Goffman discusses the use of the telephone in public settings. In one example he discusses what might be seen as a perceptually co-present interlocutor:

In public, we are allowed to become fairly deeply involved in talk with others we are with, providing this does not lead us to block traffic or intrude on the sound preserve of others; presumably our capacity to share talk with one other implies we are able to share it with others who see us talking. So, too, we can conduct a conversation aloud over an unboothed street phone while either turning our back to the flow of pedestrian traffic or watching it in an abstracted way, without the words being though improper; for even though our coparticipant is not visually present, a natural one can be taken to exist, and an accounting is available as to where, cognitively speaking, we have gone, and, moreover, that this "where" is a familiar place to which we could be duly recalled should events warrant. (Goffman 1981, p. 86)

The concept of an "unboothed" phone is a wonderful foreshadowing of the mobile phone interaction. Goffman's analysis of the body gestures, the ability of others to understand our situation as being engaged in a remote interaction, and the ability to be recalled into the physically co-present anticipates the development of mobile telephony. Indeed, it can be suggested that people have been re-boothing telephony in several ways (hand over mouth, texting, ducking into an alcove, and so on).

Goffman mentions a person's jokingly using a toy telephone that had been deposited in a trash can as a "sight gag." The person carried out a false conversation as though they he was on an actual telephone. From our remove, being used to mobile phone conversations in every conceivable place, the spectacle of a person talking on a phone in a public space loses some of its comic potential. However, for Goffman and his contemporaries the use of a telephone in a public setting was a joke in itself, since telephony was mostly reserved for private "back-stage" settings such as the office or the home. When, in addition, the telephone in question was a patently false device, it was easily interpreted as a joke. In addition, the fact that the conversation was publicly available tested the public tolerance for public interaction, particularly in that pre-mobile-phone era. To openly talk on a phone, aside from in a telephone booth, was unusual and added to the farce (Goffman 1981, p. 86, n. 6).

Though the bulk of Goffman's work focuses on literally co-present interaction, there seems to be some suggestion that mediated interaction can also be seen as a form of ritual interaction. Mediated interaction can potentially have the some of the same integrating functions as traditional face-to-face dealings.

Conclusion

The work of Goffman allows us to see the mundane, everyday form of ritual interaction. As with Durkheim, ritual interaction facilitates the social cohesion. Goffman (1967, p. 91) examines how the individual rather than the totem is symbolically charged through ritual interaction. There is a disturbing question, however. When reading Durkheim, one senses that he is looking at the origin of cohesion via the use of ritual—Turner's (1995) liminal interaction. When reading Goffman, one has more of a sense that one is reading about its maintenance. Durkheim really is talking about how when one is in an intense social gathering one becomes a part of the flow of events. At a rock concert, a soccer match, a political rally, or a meeting of a parent-teacher association, one becomes engaged in the flow of the events and also pays heed to—and joins—the mass. One sees one's friends clapping in unison to the music, cheering on the team, applauding the politician, or discussing cuts in the lunch program. As we do this, we are also aware of those around us, those people we know in other contexts, who are also behaving in the same way. They are willing to pay heed to the same things that we find important. They are willing to make public their attitudes toward a point that is also dear to us. This gives us the slight hint that they are also our colleagues and comrades. If we see them supporting our perspective, the interaction gives us an additional link in the web of social solidarity. It may increase our trust in them, and we know that we have a new theme which we can develop in other forms of interaction.

We might also find out that those who we thought were supporting our perspective are rooting for the other team or are in favor of an alternative point. Again, this helps us to place them in our consciousness. We know, in some little way, where to place them and what to expect from them. It might be that their support of the other team or the alternative political perspective makes them an example of what not to be and that they become a kind of rallying point against a broader perspective. Indeed the feeling for or against a team, a political figure, or a position on school lunches might take on a passionate edge and become a major aspect of our self-identity and a main point in our connection to others. Regardless, in this kind of interaction it is easy to see that solidarity is generated. In the heat of the interaction, the bond—or the clash—between individuals becomes obvious. Goffman obviously plays on this. He shows us how, at a much more basic level, we use these bonds in our ongoing interaction. In Goffman there is not as much discussion regarding the generation of bonds, but there is a lot of attention paid to their maintenance. Goffman reported on ongoing situations. The lives of the crofters on Unst in the Shetlands and the residents at St. Elizabeth's hospital in Washington were largely set when he arrived on the scene. We do not read of the local gatherings, the celebrations, and the festivities of these people. Instead, we learn about how they greet their guests or how they construct their identity kits. We see how they develop makeshift ways of differentiating themselves from the others. In short, we see the ongoing stuff of daily life. In this process we are presented with the repertoire of devices they have at their disposal that are used to pay the appropriate heed to others. We are not necessarily presented with the generation of these devices. This obviously folds back into the coming discussion of mobile communication. Perhaps it tells us that solidarity is developed and maintained in turns. The foundation of an ongoing relationship arises in intense co-present events and is, at least partially, sustained by telephonic interaction. This insight suggests, for example, that we first came into contact with a person at some co-present event. As the friendship develops, we maintain it through both co-present and mediated interaction.

5

Collins and Ritual Interaction Chains

The present-day theorist Randall Collins is, in many ways, responsible for the renaissance of interest in Durkheim, at least in his analysis of small-group situations. Collins is also the person who states most explicitly that ritual solidarity is achieved in co-present situations. This is bad news for those of us who are rooting for the mobile telephone. Collins adds several important issues to the mix however. He refines the idea of ritual to include insights into failed rituals. He encourages us to look at power as an aspect of ritual interaction, and he draws attention to some of the psychological dimensions of such interaction.

Collins began his post-graduate studies at Stanford University in the early 1960s, then transferred to the University of California at Berkeley in 1964. This, it turns out, was an portentous time to be at Berkeley. Collins writes that he was involved in various activities surrounding the Free Speech movement. At Berkeley he also came into contact with Erving Goffman, who informed his work on conflict sociology and provided a background for his work on interaction ritual. Indeed, Collins's term "interaction ritual chains" recalls the title of an earlier collection of Goffman essays, *Interaction Ritual*.

While Collins adopts much of both Durkheim's and Goffman's theoretical stance, his method is firmly in the camp of Durkheim. On occasion he does collect primary material.[1] However, many of his major works are based on secondary material and archival matter. In his ambitious work *The Sociology of Philosophies* (1998), for example, he uses historical archives and secondary material to trace the development and changes in various philosophical traditions (Maclean and Yocom 2000; Smardon 2005).[2]

The notion of interaction ritual chains, introduced in *The Sociology of Philosophies*, was more completely developed in Collins's book *Interaction*

Ritual Chains (2004a). According to Fine (2005), *Interaction Ritual Chains* is a "serious and essential attempt to connect the Goffmanian and Durkheimian projects so as to make clear their relevance to the understanding of the continuity and stability of social structure. . . . Collins argues that interaction rituals produce emotional energy, the gathering of which is a central motivating force for individuals. Affect is the engine of social order. Those interaction rituals that are most effective in generating emotional energy are the ones that bolster institutional stability. We seek emotional energy the way that felines seek catnip—it gives us a buzz."

The work of Collins has indeed been seen as a rejuvenation of Durkheim. According to Rawls (1996), Collins's focus on Durkheim's work *The Elementary Forms* has helped scholars to see the meso and the micro dimensions of Durkheim's work.

Interaction Ritual Chains

Goffman and Durkheim provide a theoretical legacy for Collins's analysis of the ritual and symbols in the micro-social situation. In his development of ritual interaction chains, Collins forms these insights into a comprehensive examination of how rituals affect the development and maintenance of social symbols at the micro level. Collins (2004a, p. 7) defines ritual as "a mechanism of mutually focused emotion and attention producing a momentarily shared reality which thereby generates solidarity and symbols of group membership."

The interaction ritual includes the following elements:

· two or more people physically assembled
· boundaries to outsides
· a common focus of attention through which the participants become "mutually aware of each other's focus of attention"
· sharing of a common mood.[3]

According to Collins, the sense of mutual attention plays on itself in order to intensify the shared mood:

I suggest that this [common focus] is what makes the difference between situations in which emotional contagion and all other aspects of rhythmic entrainment build to high levels, and those in which they reach only low levels or fail completely. This is above all what rituals do: by shaping assembly, boundaries to the outside, the physical arrangement of the place, by choreographing actions and directing attention to common targets, the ritual focuses everyone's attention on the same thing and makes each one aware that they are doing so. (Collins 2004a, pp. 76–77)

The operative point is the sharing of a mutual focus of attention. Collins describes the intensification of the mood through rhythmic entrainment.[4] If the grouping only comes to the point of having stereotypical patterned interactions or holds themselves to simple formalities, there is no sense that solidarity derives from the interaction; rather, the pre-existing solidarity may be consumed.[5]

However, the interaction can take the turn of developing a mutual focus of attention and a shared mood. This leads, in turn to some form of collective engrossment. The ritual outcomes include group solidarity, the sense of emotional energy on the part of the individuals, the development of markers of the shared relationship, and standards of morality (the sense of "rightness" associated with being a part of the group and the willingness to defend it against transgressors) (Collins 2004a, p. 48).

Engrossment (in Collins's terminology, entrainment) contributes to the solidarity of the participants. In some greater or smaller way, the interaction contributes to the revitalization of the individual's group identity and honors what is sacred. This may be at the very general level (as when the exercise of a courtesy to a stranger reconfirms our mutual sense of civil society), or it may be renewal of our sense of our primary group. As members of a formal group, we come together for periodic meetings. We can, for example, have a laugh with a friend, or share a bit of gossip. As with other forms of interaction, there is a ritual format that marks the telling a joke or a bit of gossip. The situation and the form of presentation can include using well-rehearsed phrases to mark the event (e.g., "Did you hear the one about the chicken?" or "Boy, do I have something to tell you! Have you heard about Tom and Mary?"). This is a ritual interaction at a very specific level with close and trusted friends. As we will see in the later chapters, I assert that these forms of interaction can be telephonic as well as face to face.

■

Like Goffman, Collins focuses on the situation. In his interview with Maclean and Yocom (2000), he says:

The way I've come to work is to emphasize that the real unit of analysis is the situation, and situations have their dynamics and individuals get constructed out of those situations. And then you're able to bring in a lot of really good

micro-research. You know, ethnomethodology is often hard to translate into anything else, but if you look at it as being concerned with "how do situations operate," for example, "how do conversations operate"—for example, what Sacks
and Schegloff call the rules of turn-taking in conversations—you can see that's
really a kind of Durkheimian ritual of maintaining this very fine-grained solidarity on the micro level. I think that helps solve a lot of problems. Encounters do
have this sort of magnetism that pulls people in.

For Collins the situation necessarily includes the physical assembly of
the group, often with some form of barrier to outsiders. It is within the
context of the situation that rituals take place. Situations are the basic
frame around a focused encounter. While permeable, certain boundaries
must be maintained. Without these, the happening will not be considered
a comprehensive unit.[6] One thing that helps to define a situation is the
sense of cooperation on the part the participants. A situation implies at
least the momentary mutual focus of attention as well as an accounting
for the broader situational reality (Collins 2004a, p. 24). If the criteria of
a focused situation are met, individuals are "swept along," to a greater
or a lesser degree, in the flux of a situation. This also implies a certain
kind of membership. There are those who are included in a situation—
by invitation, or by simple positioning. There are also those we exclude,
either by turning our backs on them or by ignoring their contributions.
Boundaries can be architecturally based (only those present in the drawing room get to participate) or introduced by the host (name cards at
the dinner table). Boundaries can also be ideologically based, such as the
exclusion of certain types or classes of persons from social situations.
Obviously, mobile communication often assumes bounded interaction, in
that two persons are naturally included in the conversation circle, though
others may inadvertently be exposed to it against their will.

In a situation, we take a line of action. Having done so, we are to some
degree constrained to maintain that line. Others who are also in the situation are similarly constrained to follow the line of the situation. We need
to pay attention to the situation and beyond that participants have a collective responsibility for the successful completion of the interaction. We
pay attention to, and indeed are committed to, sustaining the focus of the
interaction. We must play our role and hold up our end of the conversation. We must contribute to the shared sense of the event with the proper
verve—not too much, not too little. If things begin to go awry—for example, if the host spills the hors d'oeuvres—all participants have a duty to

bring things back on track. This can be done through active intervention, as when one dinner guest sees that another is about to tip over a water glass and adroitly reaches across the table to prevent that. It can also be done by casting a blind eye when participants are slightly askew in dress or performance. In any case, we see that there is a common commitment to maintaining the situation. This said, most situations allow for a certain deviation and spontaneity. We can, for example pursue other interests and lines of activity—for example, talking on a mobile phone. These side activities are done, however, only at some risk to the situation.

There are also power relationships being worked out in ritual social interaction. A boss or a teacher can demand our attention, a more powerful person can set the agenda, a man in jacket and tie can demand deference from a doorman, and so on. At the same time, the individuals involved in an interaction feel a common duty to carry off a successful engagement. This means that side engagements are risky. Special permission must be sought, or special understanding extended, if we are to do this with composure. This often involves particular interactions that mark the boundaries of the side engagement in which the individual begs the indulgence of the broader group. In this way we can see the power relationships—who requests, who grants, and who polices the side engagements—and we can also see the impingements (Goffman 1963a, p. 98). Indeed, the impulsive nature of the mobile phone has given us insight into how ritual interaction can be breached in ever more innovative ways.

The Question of Co-Presence

An essential point for Collins—and a point with which I quibble—is his resolute focus on the co-presence of the persons participating in the ritual interaction. As was noted above, Collins sets this as one of the criteria of ritual interaction. As we have seen, the focus in the work of Durkheim and Goffman is on co-present situations. In the case of Durkheim, this is almost to be expected. His empirical material came from Aboriginal ritual gatherings in the late nineteenth century, and thus it is slightly absurd to think of mediated interaction. Goffman focused on social interaction in the mid-twentieth-century United States (and in the Shetland Islands). Though he did not focus on mediation, he at least left the door ajar to the idea that it could be examined in terms of its ritual content.[7]

In Collins's estimation, the physical assembly of the group is a requisite element in ritual interaction. The groupings can be large, or, following from Goffman, they can be quite small and transient. Collins, however, is quite firm in insisting that physical co-presence is a necessary element in the development of ritual interaction.

It is easy to see how solidarity is developed when a number of people come together, fix their attention on the same theme or purpose, and allow themselves to be transported into the flux of the event. It is easy to see how solidarity arises from people "shouting the same cry, saying the same words, and performing the same action in regard to the same object" (Durkheim 1995, pp. 231–232).

Collins asserts that it is through face-to-face interactions that we have shared emotional energy and the symbolic meanings that people carry with them through various chains of interaction:

As the persons become more tightly focused on their common activity, more aware of what each other are doing and feeling, and more aware of each other's awareness, they experience their shared emotion more intensely, as it comes to dominate their awareness. Members of a cheering crowd become more enthusiastic, just as participants at a religious service become more respectful and solemn, or at the funeral become more sorrowful, than before they began. It is the same on the small-scale level of a conversation; as the interaction becomes more engrossing, participants get caught up in the rhythm and mood of the talk. (Collins 2004a, p. 48)

Based on his massive analysis of academic traditions, Collins (1998) says that it is only in co-present situations that we find the richness of interaction that allows for the development of ritual interaction. As examples he evokes the work of Sudnow and the finely interwoven turn taking in a parting. The interaction of the intonation, barely perceptible pauses between the interlocutors (indeed they are less than 0.1 second), and well-rehearsed lines such as "See you later" / "Yeah, OK" / "Bye"/ "Yeah, bye" (Collins 2004a, p. 78) shows how finely wrought our ritual interactions can be.

Collins suggests that the rhythms of co-present interaction are more easily generated and that it is simply easier to monitor the emotions, behaviors, and feelings of other participants in co-present situations. Further, it is in the richness and entrainment of co-present interaction that social solidarity is given room to develop. It is indeed meaningless to argue against the idea that co-present interaction is a highly efficient way to develop a sense of the group. Being in the same space, focused on the same event, and in the same train of thought is certainly a way to break down the barriers between individuals.

As was noted above, Collins is quite explicit in his exclusion of mediated interaction when it comes to the use of ritual. He goes beyond the perhaps understandable omission of Durkheim and Goffman to the unequivocal rejection of mediated interaction as a bearer of ritual interaction. Discussing email, he writes:

> There is little or no buildup of focus of attention in reading e-mail, no paralinguistic background signals of mutual engrossment. A written message may attempt to describe an emotion, or cause one; but it seems rare that email is used for this purpose. A hypothesis is that the closer the flow of emails is to real conversation exchange, the more possibility of a sense of collective entrainment, as in a rapid exchange of emails in a period of minutes or seconds. However, even here it is dubious that strong feelings of solidarity can be built up, or the charging of a symbol with collective significance.
>
> My main hypothesis is to the contrary: the tendency to drop ceremonious forms in email—greetings, addressing the target by name, departing salutations—implies lowering of solidarity. Email settles into bare utilitarian communication, degrading relations, precisely because it drops the ritual aspects. (Collins 2004a, p. 63)

According to Collins, mutually focused interaction and a fairly high level of engrossment lead to solidarity. There is the sense that the participants have to be drawn into the situation and that they have to invest effort to maintain the integrity of the interaction. In the case of conversation, for example, there is the need for both parties to keep the topic rolling. There can be shifts and changes in the things that are discussed, but the common willingness to maintain the conversation illustrates reciprocal engrossment.

It is the insistence on co-presence that gives the student of mediated interaction pause. Thus, in spite of Collins's ideas about co-presence, it is also possible to consider how the same mood can be engendered or perhaps revived via mediated interaction. At least with the case of mobile communication, the sense of the interaction might be broadened. Rather than an isolated, one-off event, mobile communication allows the most intense users to keep an interaction alive across time and space (Licoppe 2004; Ito 2005b).

The dynamics of a telephone call, and indeed those of a mobile phone call, can be a way to organize groupings (usually dyads) wherein there can be ritual interaction. We can tell jokes, use texting argot, and gossip, all of which are ritual forms of interaction. We can re-live previous co-present interactions and plan new ones. Indeed, mediated interaction can have a role in the development and maintenance of social cohesion. We can become entrained in the conversation or interaction with another person—or persons—when talking on the phone or in an "instant messaging" session (Baron 2004). The call or session can have a ritual quality in itself or can be an extension of other previous or anticipated co-present situations. We can tell a joke to a friend and thereby further cement our relationship. We can flirt over the phone, and we can argue. We can share the events of daily life with our spouse, or we can go through a shopping list. We can call a love object in anticipation of a date, and afterward we can re-live its high points (Ito 2005b; Prøitz 2005).

The mobile telephone, particularly in the case of text messages, can fit into the cracks and chinks of another co-present interaction. Teens can send messages to each other between (and indeed during class), thus maintaining the social link between "face sessions." When there is a pause between speakers at a meeting, we can quickly send a message to our children to remember their music lesson. We can use a mobile phone in the unoccupied minutes surrounding other more pressing events. We do not need the static reminder of the "bear clan" totem, since the reminders of our social connections are perpetually there.

Failed Rituals

An essential insight provided by Collins is that rituals can fail. A successful ritual can engender solidarity and revitalize cohesion, but a ritual carried out ineffectively can fail and thus destroy solidarity. Goffman speaks of when we are caught off base in various situations and how our "face" is threatened.[8] Indeed, Goffman's analysis of embarrassment (1967, pp. 97–112) draws attention to this issue. Indeed, the failure of a group ritual, be it co-present or mediated, can threaten the interaction of the group (ibid., p. 34; see also Scheff 1990, p. 7).

Collins points out that the strength of customs is often observed in their breach. The degree to which a ritual is successful depends on the situation and the degree to which the participants buy into the process. At the same time, people are, to one degree or another, repulsed by failed rituals. Thus, the failure to carry off a group ritual is not simply a missed opportunity; it is a negative mark with regards further interactions.

We can see the failure of ritual, for example, when we flub a greeting. It might be that some obstacle to shaking hands (such as a cluttered or crowded meeting room) gets in the way. More pointedly, we may openly refuse to greet a person who is shunned. In these situations, we note the need for the ritual and we also note its poor execution. We see the failure of ritual when a joke falls flat or is embarrassingly off key. We see the failure of ritual when our tidbit of gossip is either old news or (perhaps even more so) when the gossip is more revealing than we wished to be. ("Oh! Is he *your* husband? I'm sorry, I didn't realize that. I really shouldn't have said anything.") Large-scale events also can fail. A baseball game can be boring, the singer at a rock concert can be sick, or Uncle George can be drunk at the Christmas dinner. As with the smaller interpersonal rituals, they can be seen as failures. In these cases, the failure not only creates an awkward situation, but it can reduce the social cohesion that has been built up in other previous interactions.

These situations are, however, not without meaning. Collins (2004a, p. 51) notes that "unsuccessful rituals are important substantively as well, for every social encounter of everyday life from the most minor up to the major public gathering is to be put into the scale and weighted against the standard of ritual intensity."

The success of rituals can also reveal the contours of power. The more powerful person can often determine the limits of acceptable humor and gossip, or can decide who will be greeted and in what order. Thus, by studying the success and failure of these interactions we can expose the power structure of the situation.

Even when not omitted, everyday rituals can be done incorrectly. They can be too familiar or too distant. They can employ the wrong sequence or be used in the wrong context (for example, addressing a lover as "dear" in public).[9] Thus, rituals can fail or they can succeed at different levels of intensity (Collins 2004a, p. 15). Collins's notion of failed rituals indicates that social rituals are not simply an empty social form. The idea of the failed ritual is also important, since it provides us with a tool for identifying these daily rituals.

Conclusion

It is clear from the work of Collins that ritual interactions can range from large stadiums full of soccer fans to dyads. If mediated interaction is a part of the scene, rituals can range from a minor telephone call to an experience of the Burning Man festival and its frenzied finale.

The application of Collins's analysis to the larger and more intense forms of ritual is near at hand. Collins describes rituals that perhaps start quietly but build and build as the focus of the participants comes to be shared, the intensity increases, and the frenzy of the moment overtakes the individuals. In the process of participation, the individual loses his or her sense of self and becomes a part of the whole. This is the prototypical Durkheimian ritual. It is the forge upon which broader solidarity is minted. These events are also quite rare, and the astute reader will notice that the mobile phone has only a marginal—if somewhat nerdy—role in these events (Watkins 2005).[10]

At the same time, we find in Collins the Goffmanian line. It is not necessarily the ecstatic participation in a bacchanal that forges the ritual, but the glacial socialization into the customs and behaviors of a culture. There are events that are of particular importance in this process. These can include instruction from parents and teachers. Goffman notes that episodes of embarrassment are also effective, if painful, socialization. It is in these minor daily interactions that we learn proper greetings; we learn the importance of tone and tact. We learn to manage with composure the routine fleeting interaction, such as paying a toll or maneuvering a shopping cart through a crowded store.

Collins's notion of interaction ritual chains helps to integrate the work of Durkheim and Goffman into a model that can be applied to commonplace situations. The model helps us to understand the role of both large-scale and micro-scale interactions in the ongoing process of daily life. It is a lens through which these seemingly mundane interactions help us to celebrate our connection to the broader social sphere. It is through the rehearsal of these interactions that we practice Goffman's deference and demeanor. It is through these micro-interactions that the significance of artifacts is revitalized, and it is through our use of these interactions that we recognize the individual as a significant element in focused ritual interaction. Collins also shows how the effervescence of larger events is an important aspect of social cohesion.

Like Goffman, Collins sees ritual as the essential adhesive of society. We have our obligations to others and to the situation. It is in the co-present that many of these practices are minted. However, they are developed, maintained, and expanded in mediated interaction. As we will see below, mobile communication often tightens the social bonds of small social groups.

6

Ritual as a Catalytic Event

This book is premised on the idea that mobile communication allows for the execution of rituals that in turn result in social solidarity. As I will develop below, the mobile telephone does allow for this in some cases, while it perhaps hinders it in others. In Collins's framework, it facilitates the success of some ritual interactions (joking, banter, gossip, flirting) while hindering the success of other ritual interactions (the mobile phone as a secondary engagement). To work through this, it is necessary to summarize the preceding material.

As I developed above, the basic elements of a ritual are the mutual focus of a circle of participants and the engendering of a common mood. Further, there is generally a barrier to those who are not a part of the group. This results in a kind of entrainment where there is engagement in the situation but, more importantly, there is mutual understanding of the others' engagement. Indeed, this mutual understanding is the key.

The idea that interpersonal interaction in the context of a ritual encounter results in cohesion is a kind of syllogism. Suppose that Jens is enthralled by the spectacle of an event, and so is Rick. This is not enough to cause the development of cohesion. Cohesion arises out of Jens's recognizing that Rick is also enthralled by the event and Rick's recognizing the same in Jens. Both are individually engaged in the event, and each recognizes that the other is also similarly enthralled. It is the sharing of a mood and the mutual recognition of being engaged in a common situation that is the basis of ritual cohesion. The situation only develops cohesion when there is the catalytic effect of the ritual. It is the ritual (or the performance of ritualized interactions) that provides the mutual bond for Jens and Rick.[1]

It is the situation that has a catalytic effect. It is not simply a group of people being thrown together that causes the solidarity to develop; rather, it is people being together in the context of an entraining situation. There is the need for the participants to experience some sort of common process. The entrainment of the different persons, however, entails a separate process for each individual in that each person brings their separate biographies and social perspective into the ritual situation. They become immersed in the flow of events for their separate reasons. Jens may like the drummer playing in the band, Rick may be particularly taken with the lyrics of the songs, and Cathy may be a fan of the keyboard player. The key is that participants become mutually aware of each other's engagement. The individual's entrainment and the mutual recognition of others' having given themselves over to the situation is the basis of the cohesive link. This is, at its core, a theory of cohesion.

∎

Up to this point, I have used the illustration of a small group of people as they go through the process of developing a cohesive link in a small-scale interaction. The development of cohesion is not, however, limited to only two or three people. It is limited by the ability to recognize and play on mutual engagement. Mutually recognized entrainment can tie many people together in the context of an event.

There is clearly an upper limit as to the number of people whom can observe and share in each others' engagement. We cannot personally see, for example, the engagement of several thousand others at a large sports event or a large-scale religious service. In the case of large-scale spectacle, however, the very mass of people serves as a backdrop designed in part to inspire awe in the individual. It is not the intention that the individual will bond with all the others there. Rather, the awe of being in a large group of people contributes to the effervescence of the situation. Given this setting, the small group of Jens, Rick, and Cathy will have the opportunity to develop a sense of cohesion with a smaller group of those with who they are in more direct contact.

In small-scale interactions—such as those for which mobile phones are typically used—the notion of ritual as catalyst also applies, though in a less intense form. The types of verbal interactions carried out (greetings, partings, banter, the exchange of chitchat, gossip, etc.) are successful when there is mutual engagement. Indeed, when one party assumes that the other is drawn into the mood of the conversation but that is in fact not the case, it leads to least an awkward situation and perhaps a breach in the relationship. Thus, although there is a difference in power, it is possible to see the catalytic elements in both the large-scale Durkheimian rituals and the smaller-scale Goffmanian interaction rituals. The distinction between large-scale and small-scale rituals is, however, in need of clarification, particularly since mobile communication fits better in the latter.

The Careers of Ritual Solidarity

The works of Durkheim and Goffman also suggest that there are different "career paths" for ritual interaction. For both Durkheim and Goffman, the ritual involves a mutual focus, a common mood, and a bounded group. Both also see the ritual as a way to generate or to play on cohesion in social interaction. There are, however, several differences. These involve, among other things, the scale of the ritual, its periodicity, and the role of the participants. If the idea of ritual interaction is to be applied to the interpersonal interaction carried out via mobile phone, it is necessary to examine these issues.

Durkheim's work evokes the idea of a liminal process in which some large-scale event is the catalyst for bonding between individuals. In contrast, Goffman does not describe large-scale events, but rather small-scale interpersonal interactions. There is the sense, then, that Durkheim's approach is useful for describing the situations in which cohesion is engendered,[2] whereas in Goffman's work we see it being used in mundane events. If we think about the career of ritual entrainment, Durkheim describes its development and Goffman its use in practice. Collins helps us to see both of these contexts. Indeed they are particularly obvious in the case of failed rituals (Collins 2004a).

It is tempting to see Durkheim's notion of ritual as a set of intense liminal ritual celebrations where the group comes together, develops the emotion and intensity of the particular event, recharges the importance of the totem, and then moves into the mode of everyday, non-ritual life. (See figure.) The level of social cohesion varies with the cycle of the ritual interactions (Durkheim 1954, p. 130). The potency of the totem is also associated with the cycle of ritual events. If the totem is not occasionally

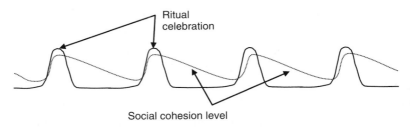

Episodic ritual celebration and social cohesion in Durkheim.

recharged through being the focus of a ritual process, it will not have the same symbolic weight in the eyes of the followers.

The Durkheimian approach also applies to staged events. There is a basic division between those who are responsible for a staged event's production and those who are seen as the event's normal participants. These are two relatively well-defined and different spheres of activity. Those who are staging the event are responsible for sequencing it so as to engineer the proper mood. They also have a relatively powerful position (Couldry 2003; Bourdieu 1991). A rite of passage, a large religious ceremony in a football stadium, a rock concert, a cotillion for the upper class, or a basketball game involves a corps of individuals who stage the event and encourage the general participants to adopt a mood and a line of behavior (for example, singing either hymns or "fight" songs, waving arms in time with the music while holding lighters aloft, repeating a prayer, proposing a toast, or participating in a "wave"). In a large-scale ritual, it is not the responsibility of the general participants to generate the setting or to control the mood-setting machinery. That is the job of the persons staging the event.[3] In the large-scale ritual, devices and allurements are assembled to engender the appropriate context in which, among other things, cohesion can take place.[4] The role of the normal participants is to be transported by the event. If this is successful, the participants adopt a similar mood and become available for the cohesion processes described in catalytic rituals.[5]

In Durkheim, then, there is a clear idea of how solidarity is generated, but the actual fate of the cohesion between the ritual events is not as well developed. Thus, this is not necessarily the most fruitful area of inquiry for the analysis of mobile telephony. It is not in this context that we can see the role of mediated interaction being played out.

■

In contrast with large-scale Durkheimian rituals, Goffman (1967) focuses on smaller, incidental situations and their consequences for social interaction. As opposed to the episodic events described by Durkheim, Goffman paints a picture of more or less constant interaction ritual. We are almost constantly greeting somebody, taking our leave, telling a joke, exchanging some gossip, and moving in and out of the flow of supping, speaking, soothing, or serenading. In each case, we draw on props and guises that have been mutually constructed in previous interactions and we invest our willingness to engender a mutual focus and mood in the interaction. The event may be as fleeting as the payment for a cola at the counter or as deeply engrossing as flirting or telling a joke. According to Goffman, our sense of self and our notion of others are played out in these interactions.

In a small-scale Goffmanian ritual, there is the use of well-oiled routines. As we are socialized, we learn to read situations and to draw upon the devices that are appropriate then and there. These interactions are not deeply liminal in the sense that we are transported by a particularly good greeting sequence; rather they are perhaps more along the lines of habitual interaction. In the case of a well-done interaction, there can be the enhancement of interpersonal solidarity. It can add to the positive equilibrium of the group. If done poorly, the interaction can have a kind of centripetal force that convinces some participants that this is not the group for them.

In a small-scale ritual, it is the job of the participants to not only fall into the mood of the session, but to also participate in engineering it. Unlike the large-scale ritual, there is not necessarily a division of responsibility between those who stage the event and those who participate. We rely to a far greater degree on our own devices. Instead of relying on the sponsors of an event to provide the appropriate ancillary effects as might be expected in a large-scale ritual (background music, lighting, rhythmic chanting, etc.), the atmospherics are the responsibility of the participants (the warm smile you give when first greeting the other, the large-eyed facial expression when you hear the latest gossip about her ex, the openness to bantering with the others at the table, etc.). The point is that the people who are offering the greeting, telling the joke, or dishing the dirt are more squarely the architects of the event. We use the routines we have learned from similar situations. We use our reservoir of ploys and gambits in order to successfully extend a greeting, tell a joke, use some argot, or exchange some gossip.

Small-scale rituals are more limited in scope, but they are far more frequent. They take little time to complete, and we are stumbling into them continually. For every large periodic sporting event, wedding, rite of passage, or religious observation in which we participate, we engage in hundreds if not thousands of everyday mundane greetings, partings, conversations, and interactions.

As has been noted, Durkheim's work provides insight into the genesis of cohesion, but not into how it is played out. There is the assumption in Goffman that all the various props and effects are in place when we come onto the scene. Absent is Durkheimian discussion regarding where they come from and how it is that we have agreed on their meaning.

The assertion is that the forms of social interaction are a kind of incremental residue that is developed over time. To shift metaphors, with each iteration, there is the production of some "baggage" that pre-forms future interactions. In the Durkheimian ritual, the forms of interaction can become formalized. For example, often there is the same liturgy in each event. In Goffmanian interaction, the particular form of interaction is not set, but rather it draws on a repertoire of devices and patterns that have been collectively developed over time.

The understanding of the routines comes from practice. As we are exposed to different situations, we work out the best way to exchange greetings or to tell a joke with the other people involved in the interaction. Indeed, a part of socialization is learning to read a situation and understand when an innocent joke about a chicken crossing a road is more appropriate than a less innocent joke about a traveling salesman and a farmer's daughter. As we are socialized, we learn which devices are appropriate in which context. The interaction is not catalytic; rather, it is in the form of "here we go again."

■

Collins (2004a, pp. 271–278) sets up both the Durkheimian and the Goffmanian forms of interaction as continua starting at formal rituals, passing through sociable situations, and ending in less focused "open public situations." On the one hand, he discusses formal rituals as being highly focused, scheduled, and scripted. These large-scale events are contrasted with weakly focused, unscheduled, unscripted informal rituals. Collins uses this scaling to examine how social stratification might develop as a result of our access to the different forms of ritual interaction.

The important idea that Collins brings to the discussion is that we can, by turns, be exposed to larger-scale formal interactions in which group cohesion is easily founded. At the next turn, we play on that cohesion in smaller group interaction. Finally, in a third iteration, we are again in a larger formal situation, but we can also be so concerned with the smaller-scale local interaction that the broader sweep of the formal interaction is lost on us. Thus, according to Collins, the career of ritual interaction in a group can shift from the formal scripted forms into less formal but equally intense types of interaction.

In anticipation of my extension of Collins into the mediated sphere, it is possible to see how small-scale mediated rituals between individuals can, for example, serve as a prelude and perhaps a resolution to major Durkheimian events. In most cases there is either the one or the other, but in some cases mobile communication has mapped these onto each other— for example, teens texting one another before and after a date (Ito and Okabe 2006), texting one another at concerts (Grinter and Eldridge 2001), or cobbling together various communication systems in order to broadcast and comment on rituals as they unfold (Watkins 2005). It is the mobile telephone that allows the extension of interaction beyond the physical space of an event. The result of this may be the reduction in the common focus of the co-present event itself, but it may mean that the informal group that is exchanging messages is more tightly integrated. These mediated interactions can also be "stand-alone" events that take place in the broader flux of everyday life.

The Boundary between Ritual and Routine

Durkheimian ritual is relatively easy to distinguish. As I have noted, Goffman took many of the same elements and applied them to common daily interaction. At the more nuanced Goffmanian end of this scale— that is, in the case of what Collins calls sociable and open public situations—it is worth considering the boundary between the routine and the ritual. Is it, for example, a ritual interaction when two people pass each other on a narrow portion of a sidewalk? The answer to this must be found in their degree of mutual engagement. Collins, like Goffman, discusses the nodding of a head or the raising of a hand as potential rituals. To be sure, a nod can speak volumes if, for example, it signifies a father's re-acceptance of a prodigal son. However, can it always be considered a ritual gesture? Pushing the definition even further: For example, when we have to carry a plate of food from a serving line to our place in a crowded cafeteria, we negotiate with other cafeteria patrons, using a series of glances, and we engage in slight re-positionings and bodily adjustments to help us as we navigate past each other. These gestures and glances are rarely misinterpreted. There can even be bodily contact between persons who do not know each other. For example, we use slight touches on the arm or shoulder to help us manage the situation and to get past the other person without too much fuss. Is this ritual interaction? The discussion of rituals is full of descriptions of sacred objects, social effervescence, and obeisance. But just how subtle can the interaction be and still be seen as ritual?

It is difficult to see where the practical boundary between ritual and routine can lie. This is, however, important when one is thinking about the types of interactions a mobile telephone makes possible. Such interactions can encompass other co-present individuals, those with whom we are talking, and even those to whom we are texting (who may not read the message for some time).

One approach is to consider the degree to which the situation brings about a mutual focus and the degree to which it can potentially engender effervescence and entrainment. It is through this kind of engagement that we reinvigorate the symbolic nature of the icons, be they physical objects or gestures. Thus, without the mutual engagement of the participants it is difficult to see how an interaction can be seen as the kind of ritual discussed here. To be sure, the level of engagement when we are maneuvering in a cafeteria is not the same as when we flirtatiously dance the samba with a partner. Nonetheless, when interacting with others in the cafeteria we are also drawing on a reservoir of signals and gestures to which we have attached meaning. We also see that when the cafeteria maneuvers fail the result can be broken china. Thus, by looking at the negative possibilities we see that there are necessarily also mutually recognized ritual interactions in the cafeteria, though at a far more nuanced level than we use when dancing the samba.

In minor daily interactions, are we contributing to a mutual form of solidarity? Are we becoming members of the bear clan, or supporters of the Nebraska Cornhuskers? Alternatively, are we, as Goffman suggests, engaged in the broader but much less intense process of supporting the solidarity of a culture? In the minor events of the day are we generating solidarity in any meaningful sense or are we merely drawing on pre-developed patterns of social intercourse that can be applied across the board? To the degree that there is a common engagement in the setting and we draw on our common sense of decorum in the cafeteria, we confirm our membership in the cafeteria society and we pay heed to the rituals of cafeteria life.

Many of the elements of Collins's ritual are in place when, for example, one is thinking of buying a cup of coffee shop. There is isolated interaction between individuals: after waiting in line, the customer places an order and pays. There is a common focus between customer and the order taker. This may take the form of a chirpy sentence such as "Good morning, what coffee product may I get started for you today?" Here we see exclusive interaction, a common focus of attention where participants are "mutually aware of each other's focus of attention," and the sharing of a common mood. But does cohesion arise from this interaction?[6] If this is a one-time interaction, the cohesion might not find any traction. However, if the customer is a repeat customer and the order taker is the same from day to day, they might generate some familiarity. Indeed, Collins suggests that the more often we repeat an interaction, and the more focused the ritual occasions, the stronger the group's boundaries (Collins 2004a, p. 273).[7]

As long as the participants in an interaction believe that they are generating solidarity, it is true in its consequences (Thomas 1931). If both participants believe there is a common mood, there must be some form of cohesion. If, however, one is truly drawn into the interaction but the other is not, there is obviously not mutual entrainment. Yet there is, at least on the part of the one who is gullible enough to be taken in by the coffee-shop employee's greeting, the sense that it is a nice day and the world is really OK. In the words of Giddens, along with a cup of coffee the customer gains the sense of ontological security in the interaction, which is quite a bargain if you think about it (Giddens 1986, p. 375). As a general rule, however, this kind of asymmetric interaction cannot be seen as a successful ritual, since there is no mutual sense of the event.

In some cases, such as the ritual aspects of a large sporting or religious event, there can be sequential development of mutually realized entrainment and sharing of a common mood. In a situation of this kind, we are building up the social solidarity of the group. But there are other rituals that operate at a much lower level of abstraction. It is not the intention that we cement an enduring relationship with these smaller interactions. They are, however, the smaller marginal interactions with which we patch together our daily lives. Interestingly, the ability to do these types of interactions is extended by the mobile phone.

Conclusion

While there can be passing engagement in both the Durkheimian and the Goffmanian form of ritual, both can also arrest our attention. We can be transformed by the excitement of the game and the roar of the others in the stadium; just as completely, we can be gripped by the gossip being told about an unfortunate neighbor.

It is the small-scale ritual that is more central to the understanding of mobile communication (Humphreys 2005; Ling 2004b). To the degree that mobile communication is akin to the small-scale interaction of everyday life—indeed, to the degree that it is becoming integral in mundane life—it provides insight into how mobile communication contributes to social cohesion. A phone call or a text message is not usually a major entraining event. Rather, with mobile communication we get the small-scale talk and the embroidery of everyday events. We get the planning and the recapping of other, perhaps major catalytic rituals. In addition, we get the exchange of endearments and insults. We get the planning of social interaction and we get the machinations of commercial activity. In short, we get the stuff of routine life.

7
Co-Present Interaction and Mobile Communication

A leitmotif in present-day society is that the mobile telephone has disturbed interaction in the local sphere. It is a device that seemingly is in dire need of discipline. In a 2005 study of attitudes among a random sample of individuals in the United States (University of Michigan 2006), 62 percent of the respondents agreed, or agreed strongly, with the statement "Using a cell phone in public is a major irritation for other people." Interestingly, this is a very age-related phenomenon. The data shows that among those aged 18–27 only about 32 percent had this attitude. This compares with 74 percent among those in the 60–68-year-old age group.[1] Rice and Katz (2003) examined some of the same issues in a national sample of the United States from 2000 and reported a widespread sense that the device had encroached on public space.[2]

The mobile phone seemingly encourages people to have the most remarkable conversations in public spaces. It provides us with a way to forget the boredom of a bus ride or a wait in a doctor's waiting room and instead interact with our best friend who is miles away. All of this has happened since the mid 1990s. Rapid adoption of the technology has meant that we have been forced to adjust our ideas of propriety in what might be called a slapdash way. The extent of the revolution can be seen in this exchange on a website that allows visitors to pose etiquette questions that are then answered by other readers.

Question: If someone is using a cell phone in the bathroom stall next to me, is it rude to flush?

Answer: That's funny! This recently happened to me and I was flabbergasted that someone would actually discuss a real estate deal while relieving themselves. I don't know, it just seems wrong somehow! Then again, I don't even own a cell phone, so what do I know . . .

I would have to say the answer is no, it's not rude to finish doing your business if someone in the next stall is using their phone. What would you do, wait until they left and then flush? Or just walk away without flushing? I think it would be more rude not to flush and walk away leaving a nice surprise for the next person.

It's assumption of risk on the cell phone user's part, imho [in my humble opinion]. If one makes/takes a call in an area that is designated for a specific purpose, it's reasonable assume the sounds of that purpose will be going on in that area, i.e. the flushing of toilets in a bathroom. Just like if the person used their phone in a noisy construction area; should the workers stop what they are doing so the person on the phone can have a conversation without all the noise?

I would say responsibility lies with the cell phone user—if they don't want to have the sounds of people using the restroom going on in the background as they use their phone, then they should consider using their phone elsewhere.

After all, is nothing sacred anymore? (Answerbag 2006)

There are several timely issues here. The most obvious is the sense of indignation that another person would talk on his or her phone in such a touchy physical location. In a sense, it is bringing the broader world into what is often seen as a sacrosanct place. Thinking about it in terms of ritual and social cohesion, the public bathroom is perhaps that area of life where we are most constrained by rules since we are, in reality, making fundamental adjustments in our facade. In order to carry off the occasion with aplomb, we need a deeply entrenched set of conventions upon which we can rely. The indignation of the person writing the response (there were several more along the same lines) shows that the telephone user has violated the rules.

Another issue here, however, is that the telephone user in the bathroom has included in the situation not only the co-present people but also, potentially, his or her telephonic partner. In the original question, the writer implies that by making an unexpected noise, indeed a noise that has perhaps a unique ability to expose the location of the conversation, the telephonic interlocutor would be given information that he or she otherwise would not like to possess. Thus, the person writing the question is more concerned about the sensibility of the person on the other end of the phone call than about the co-present telephonist.

The incident cited above illustrates the complexity of interaction when considering the introduction of mobile communication into a setting. Its use has implications for both the local setting as well as and the mediated relationship. The way we talk or text and the location in which we choose to interact influence the situation.

When making a phone call we, in essence, have two publics: the person with whom we are speaking or texting and the people with whom we are co-present. Each of them has certain rights. Each of them has the expectation of a certain level of engagement and shared disposition. In Goffmanian terms, the management of this double front stage can be complex. If too much attention is given to the co-present situation, the telephonic interaction might suffer and vice versa. The setting in which the telephonic interaction takes place can also impinge on the form of the interaction.

In addition, and in contrast to the assertions of Collins, I will explore how mediated contact via mobile telephone can also be seen in the context of ritual interaction. The way we greet one another over the phone, the way we relate stories, and the way we use the telephone to organize our daily life show that, in many respects, ritual interaction can be carried out via interactive media. Further, there are particular forms of interaction and parlance that seem to occur only via mobile phone that can be seen as mediated ritual interaction. Thus, I am interested in exploring how Collins's sense of ritual interaction can be expanded into the area of mediated communication.

In the case of mobile telephony in co-present situations, however, there are several different classes of interaction. There are situations wherein where the mobile phone (the physical object) serves as the focus of interaction, and there are situations in which the mobile phone serves as a secondary involvement. The mobile phone can also be a repository of personal history. Finally, there are situations where the use of the mobile phone is a barrier to co-present interaction.

The Mobile Phone as a "Stand-Alone" Object in Co-Present Situations

The rapid and widespread adoption of the mobile telephone has made it an object of comment. It is one of the iconic forms of our time. It is the latest item that we carry on our person in daily life, along with watches, glasses, wallets, purses, and music players. Its newness along with the technical and design developments mean that it is not a neutral item; rather, the physical form of the handset is an object of interest. When, for example, we see the latest Samsung, iPhone, or Sony Ericsson, we may ask the owner about its characteristics.

Like other mundane personal objects, mobile telephones and other devices give us insight into the style and status of those who use them—a businessman and his BlackBerry, a teenage girl and her Nokia, and so on. The object itself is actively interpreted. Its use in co-present situations is symbolically invested and can be seen in terms of its contribution to ritual interaction. At the most basic level, the mobile phone—the physical object in itself—can be a focus of co-present interaction (Taylor 2005), just as when someone plays with keys, coins, or a pencil.

The Mobile Telephone as a Physical Object

Although I argue that the constant contact afforded by the mobile phone replaces the need for the Durkheimian totem, it is possible to see the physical object of the mobile phone as a symbolically laden object—that is, as a totem. The type of device we have and the functionality of that device provide insight into our tastes, our style of consumption, and perhaps our allegiance to certain groups.

The mobile telephone itself—its style or model—can be a topic of conversation. Those who have only recently met can exchange small talk with regards the type of mobile phone that they have and the advantages or disadvantages of various technologies. It is an easy source of mutual engagement and topic that provides us with a point of entry other topics are not as near at hand. A particularly new, old, or advanced device can be an occasion for discussion in itself. The mobile phone can give us insight into the status or fashion sense of others (Fortunati et al. 2003; Ling 2001). Among some groups, it can be seen as a blessed object. It is also a conduit through which gifting rituals in the form of phatic interaction, courtship, fashions and fads, and a host of other communications can be organized.

The device can, for example, be used strategically in face-to-face interaction. Taylor (2005) found that by interjecting talk about our mobile phones into the conversation we can turn the interaction away from other, more awkward topics. He suggested that teens talk about their mobile phones as a way to organize the topics of interaction and also as a way to determine participation status in local group settings. It can be used deliberately in the interaction as a way to lead conversation away from dangerous shoals. For example, Taylor documents situations in which female respondents deliberately change the subject to talk about mobile phone terminals in order to end the more awkward discussion about boys and issues with boyfriends (ibid., pp. 150–151). By momentarily making the device into the primary engagement, they set a boundary on a particular topic and put the interaction onto a new course. Other such ruses are available, classically discussing the weather or the fate of a sports team.

Weaving of mobile telephone talk into an ongoing conversation to bridge awkward moments, to mark or comment on status, or to turn the conversation points to the fact that such ploys are a part of ongoing social interaction. If, for example, we have the sense that the conversation is heading toward difficult straits, talking about a neutral topic serves as a way to make a transition to other issues. Glancing at the telephone or quickly checking for new messages can be a "break" from the mutually focused activity of the co-present conversation. It is a breach in the mutual engagement and thus marks a minor ritual failure. However, our ability to draw on these gambits and the willingness of the other(s) to accept them shows how the device can become a prop in the ongoing obligation to maintain the specific mood of a social interaction.

The Mobile Telephone as a Repository of Personal History
In addition to be being a symbolically imbued object, the mobile telephone has evolved into a significant repository of personal information. For teens, the history of text messages, their call logs, and the number and status of the individuals in the name list on their mobile phone are all important bits of information that can be displayed to engender one effect or another (Taylor 2005; Ling 1999). As the following observation shows, it can be a repository for information needed only for a short period of time—that is, a replacement for personal reminders:

Observation In a post office in Oslo, a man approached the counter and took out his mobile phone. After recovering a certain "screen," he somewhat stiffly held out the mobile phone and showed it to the woman on the other side of the counter. He said "I want to get this." He had apparently saved the reference number to a package on his mobile phone. The woman took the device in her hand and read the information. She then placed it on the counter before moving to the rear part of the post office in search of the parcel.

This was a slight breach with the traditional interaction in a post office. The telephone in this case had replaced the more normal piece of paper as the repository of the information. The interaction between the customer and the woman in the post office, his slightly stiff movements and her leaving the device on the counter instead of taking it along with her, as she might have done were it a less expensive piece of paper, indicate that the use of the mobile phone in this capacity is not thoroughly integrated in the normal flux of interaction.

∎

A mobile phone can also be a repository of more permanent digital artifacts. Consider this observation:

Observation   In a shopping center in Oslo, two middle-aged men and a middle-aged woman were standing near an escalator. The three formed a small triangle, all easily within arms' reach of one another. They were laughing and seemed quite glad to have each other's company. One of the men took a "clamshell" mobile phone out of his shirt pocket along with his reading glasses. He put on his glasses and then opened the telephone and used the keypad a moment. He had found a picture in the memory of the telephone. He passed the phone to the other man, who looked at the picture. The first and the second man exchanged several comments before passing the phone to the woman, who also looked at the picture and added some comments before giving it back to the first man. The triad continued to talk jovially. The man with the phone put his reading glasses back on, retrieved another photograph on the mobile phone, then passed it around again. This time the phone went around more quickly with fewer comments. The man with the phone folded it and put it back into his shirt pocket along with his glasses. The three continued to talk.

Photography has an important role in society. A photograph often captures a something significant—for example, a wedding, a graduation, a grandchild, a first tooth, a new boyfriend, or a memorable vacation. The viewing of a photograph—in this case facilitated by the use of a mobile phone in its "stand-alone" phase—can bring back fond memories of an event. A photograph is often woven into our self-narratives and, as in the observation above, used to illustrate themes that arise in the flow of a conversation. It can intensify the flow of narration by giving it a concrete visual element. This is ostensibly the case of the man with the camera phone observed here. The photograph can confirm that we have participated in activities that are related to the current topic of conversation. ("You asked how my son is doing. Here is a picture of us on vacation in Spain last January.") A photograph used in this way contributes to the entrainment described by Collins. It can help to foster the discussion and facilitate its development in certain directions.

At the same time, a photograph can fail as an object of common interest. The photo may be poorly focused. The scene of the beach can be inconclusive and poorly framed. The photo of the child's first steps or first outing on a bicycle can fail to inspire the same nostalgia or excitement in a non-familial viewer that it excites in the over-proud parent. The photograph can fail to give the viewer a true sense of just how nice it was when the family was gathered for the reunion, or the erstwhile photographer can be over-eager to show a marginally willing viewer his or her entire backlog of photographs. In short, the event of viewing of the photograph can go flat.[3]

The viewing of photos can, however, contribute to the entrainment of the interaction, as in the observation reported above. What is new here is that the electronic storage of photos means that we can have a much-expanded number of photos easily available. The practice of having a photo of a child, a boyfriend, or a girlfriend can conceivably be extended far beyond the limitations imposed by photos printed on paper. It is conceivable that electronically based photos will allow us to vastly expand our ability to include illustrative photos in normal conversation. In addition, and more to the point here, the ritual of showing photographs and the resulting symbolic reverberations may be included more directly into our everyday interactions. That is, we will be able to charm or bore our co-present interlocutors with photos to an even greater degree than has been possible.

The mobile telephone also expands our ability to capture images in everyday life (Ito 2003; Ling et al. 2005). The fact that we have the device on our person and the fact that it can potentially hold many more photos that if we were to carry their paper-based equivalents mean that the mobile phone (qua camera/photo album) allows for a restructuring of local narrations. There are new possibilities to bring our external experiences into local conversation, and new possibilities to be bored by others' vacation photos.

Mobile Communication and the Co-Present Situation

I have already discussed the mobile phone (the physical object) and its role in co-present interaction. The assumption was that the device was itself passive. It is, however, a mediation technology that can be used to communicate with others. I will now examine how the mobile phone threatens to become a secondary engagement, how it can be a barrier to interaction, and how we interact with the co-present situation when it is actually the mediated interaction that is dominant (Rice and Katz 2003).

The Mobile Telephone as a Secondary Engagement

A common understanding of mobile phones is that they are what Goffman would call a secondary engagement with respect to the dominant involvement (Goffman 1963a, p. 44). A social occasion—be it co-present or a mediated interaction—obligates us to be ready to recognize certain types of involvements at any particular moment. Certain activities (flirting, an argument, a meeting) can only be carried out as primary activities. For Goffman (ibid., p. 44) these were also co-present activities. The example of the mobile telephone shows that their sanctity, however, is often under threat. We have to weigh the claims as to which particular involvement is, at any particular time, the dominant one and which involvements are subordinate. (Is it the flirting, the business meeting with an important if boring client, or the text message from our spouse that is most important?) Managing involvements can become quite complex, particularly given the fact that a mobile phone can seemingly come to life at the worst possible moment.

Quite often, secondary involvements have the function of filling in the gaps between more pressing obligations. When awaiting the start of some activities, we can easily engage in other activities that allow temporary engrossment. While there is a co-present engagement pending, or perhaps in progress, a secondary interaction can be unfolding. These must be easily collapsible when we are, for example, called upon to participate fully in the meeting or the flirting. Secondary engagements can include small talk with others who are also waiting, reading, and so on. Mobile telephony can also be seen as a subordinate engagement. The mobile phone can be used for asynchronous mobile text-based interaction while we are waiting for another activity to occur (such as the arrival of the tram or waiting for our dinner partner to arrive) or during lulls in action.

There is the question, of course, as to which ceremonies, meetings, or class sessions are seen as the dominating activity and to what degree secondary activities (such as sending and receiving text messages) can sneak in without disturbing our sense of propriety or decorum. While mobile voice telephony offers the possibility for engrossing interactions, text messaging can often take place without disturbing the gathering, be it a class or a bus ride.

Text messaging is more asynchronous than a traditional verbal conversation and thus does not lay the same claim on the interlocutor's attention as does synchronous conversation. In addition, it is discreet in that it does not include the need to speak aloud. Thus, if deftly used, text messaging is a kind of parallel interaction that does not need to be attended to in the same way as a verbal interaction. This "lightweight" form of interaction can easily be fitted into loose moments when the individual is not otherwise disposed (Ito 2005a, p. 133). This, however, has implications for the form and extent of the messages that are likely to be exchanged. Text messages are necessarily short and not necessarily socially absorbing (Hård af Segerstad 2005a; Ling 2005c). Common themes are coordination of other activities and simple greetings.

The "rules of engagement" are different with text messaging. The pauses are interpreted differently and the phrasings are not the same (Döring and Pöschl 2007). Nonetheless, a text-message interaction can be as demanding of attention as a verbal co-present interaction. Indeed, when compared to the typical small talk in which we might engage in a public situation, the content of a text-messaging interaction with a lover or a close friend may be more in tune with our core interests. Put into the terms of ritual interaction, the mutual engagement and the common mood is invested more in the mediated interaction than in the co-present one.[4] People can receive and send serious and emotive messages—for example, messages dealing with the internal dynamics of a teen relationship (Prøitz 2006). Often, however, the messages are composed and sent with the recognition that the reader may soon have to shift activities and drop the SMS-based engagement.

Though the use of the mobile phone can fit into the empty moments of other co-present interactions, it can also become the main act. This is more the case with voice calls than with text messages. Unexpected incoming calls have the potential to cut through existing social situations and to erupt onto the scene. In these circumstances, there is often the question as to what activity is primary and what activity is secondary.

When the phone rings, we need to act quickly to either disallow the call and give priority to the co-present activity or rearrange the social furniture and give the call priority. In either case, the sanctity of the co-present interaction is put to a test. The degree of engrossment is an important element here, since the alternatives of taking the call or not taking it are a concrete check on the status of the engagement.[5]

The use of a mobile telephone in a co-present situation helps us understand the sanctity of the primary engagement and its vulnerability to other secondary interactions. As an amplification of Collins's centrality of co-presence, the use of the mobile telephone as a secondary device, and more importantly our willingness to put it away when the main event starts, points to the importance of the main engagement. When we are called to order in a primary co-present activity, our willingness to stow away the secondary engagements heightens the focus of the main event. It is a sign to the others that we are available for mutual interaction. If, however, we continue to use it during other co-present interaction, it obviously drains the event of its verve.

Text messaging is a little more difficult to place in the landscape, since it can be carried out as a sly subordinate activity. In addition, text messaging is asynchronous and thus we do not have to answer incoming messages immediately. This means that text messaging can be used in the unoccupied minutes surrounding other more pressing events, or messages can be composed "under the table." The degree to which the texting or the co-present interaction is the center of engagement can have meaning for the success of the ritual interaction.

The Mobile Telephone as a Barrier to Interaction in Focused Co-Present Situations

The widespread adoption of the mobile telephone has also had the secondary consequence of creating gaps in the co-present interaction. If, for example, our co-present partner is on the phone, we must often fill in the time with other diversions. This is a particularly interesting balance since we are often within earshot of the telephone conversation. In many cases, however, we need to take on the pose of being otherwise engaged—reading a newspaper, observing other persons nearby, arranging our clothing. This engagement has to be such that it is easily packed away and the co-present interaction can be revived after the call.

In such cases, there is a kind of ritual used by the person taking the call that is associated with being excused from the co-present interaction to take the call and then another round of ritual interaction to revive the co-present interaction. Thus, the greeting and parting sequence for the actual telephone call (in itself a small ritual interaction) takes place within another ritual interaction as a kind of Russian nesting doll (Ling 2004b).

Co-present interaction does not always stop when a call is received. For example, when a person receiving a call is in a meeting, co-present interaction can continue while the call is handled.

Observation   A woman's telephone rang while she was at a meeting with three men. She looked at the screen and took the call. The interaction was along these lines:

Hi, Jenny. What is it?
(response from telephonic interlocutor)

Oh, do you have to do that?
(response from telephonic interlocutor)

Are you sure? OK, but you see I am in a meeting. I can call you in a few minutes.
(response from telephonic interlocutor)

OK, yeah.
(response from telephonic interlocutor)

I will call you in a few minutes. Don't start with that before I call OK?
(response from telephonic interlocutor)

Yeah, bye.

While she was on the phone, the woman ducked down and turned away from the meeting table. The tone of the interaction indicated she was talking to a child. The conversation continued among the three remaining meeting partners but was somewhat lame and distracted during the call. The pauses between conversation turns among the three lengthened somewhat. The "meeting dialogue" stopped slightly as the woman went through the closing sequence with her telephonic partner. The woman ended the conversation and returned to a seating position. Facing her meeting partners as she put away her phone, she commented that her children wanted to bake cookies.

In this case, the person receiving the call withdrew marginally from the local social interaction in order to carry out the telephone call. The flow of the phone conversation and her ducking back from the table indicated that the woman was trying to both address the issues of the child and minimize the impact of the call on the flow of the meeting. In this situation, she simply scooted her chair back from the table where the others were still engaged in the meeting conversation. She was still, however, within the audible range of the others, and so the two interactions were competing for dominance in the audible sphere. In this case, the meeting talk limped along and the telephonic interaction was rushed. During the closing sequence of the telephone call, the dialogue in the meeting was almost at a standstill since it signaled to the meeting participants that the woman who had received the call would soon be returning to that interaction and might need to receive a short résumé of the interaction during her "absence."

Fine-grained interaction was being carried out in two spheres. Intonation, the recognition of pat sequences, and the tempo of interaction were being used in the telephonic interaction—albeit in a slightly exaggerated sense, since the child was less able to interpret these paralinguistic cues. While overtly for the benefit of the child, they were also being interpreted by the other participants in the co-present meeting. During the whole interaction, the co-present individuals were working to maintain the "meeting" situation while being confronted with the minor distraction of the call. Nonetheless, the "meeting" interaction was hampered, and it was essentially suspended during the closing sequence of the telephone call. After the call was completed, the meeting interaction was repaired by providing the woman with a short summary; then the meeting continued. The participants were refocused on the co-present interaction. The outside distraction of the call was put to rest and the mutual forms of meeting etiquette were again in place. The woman dropped her mothering persona and re-entered her meeting persona.

When describing this it is interesting to draw on Goffman's (1971) sense of civil inattention. Goffman, however, describes civil inattention as something that is dispensed by the "viewer" as a kind of courtesy to the person who is somehow out slightly of sync with the general drift of the gathering. The sense is often that a single person among many is in violation of the situation. Civil inattention is a mechanism that allows us to proceed with the situation even though something is slightly amiss and one (at least) of the participants is not properly attending to the situation. Civil inattention encapsulates the breaching partner and allows the partners who respect the co-present scene to continue with their participation in the situation.

Seen in terms of ritual, the use of civil inattention means that the main engagement is being threatened and that the mutual focus of the participants is somehow at risk. If the infraction is minor, or if it is a legitimate distraction, the others "turn a blind eye" to the event and carry on as best they can. When the specific distraction is over there may be some way of re-admitting the errant partner back into the flow of the dominant event. It might include some simple dialogue (such as "Carrie, we were thinking that the project meeting should be on Thursday") updating the previously "absent" partner. In other cases, it might simply be the quickening of the dialogue in the dominant discussion.

In the preceding example, the telephone conversation took place within the context of an ongoing interaction. It is also possible to examine how the telephonic and the co-present overlap one another. Consider this observation:

Observation   A middle-aged woman standing outside a London subway station was talking on her mobile phone. Another woman came out of the subway station and the two waved to each other as if to catch each other's attention in the crowd and also to give and receive a greeting. The woman talking on the telephone continued to talk for the 10–15 seconds it took the other woman to cross from the station entrance. When the two came together they embraced while the woman on the telephone continued her conversation. The woman who was on the telephone told the other woman the name of her telephonic interlocutor and then held the phone to the ear of the newly arrived woman for a couple of seconds. She then took the phone back and completed her farewells and turned her attention fully to the newly arrived woman, and the two walked off into the crowd.

The woman on the telephone was managing both a greeting and a farewell simultaneously. Perhaps the telephone call was being made as a kind of "filler" while she was awaiting the arrival of the other friend. The arrival of the other friend from the subway station gave the telephonist the need to complete the call. However, since all three partners seemingly knew each other, the telephonist took the opportunity to at least give the other two (her telephonic interlocutor and the co-present partner) the opportunity to say a quick hello before turning her attention from the one to the other. Thus, the woman on the telephone was, in one sense, neither here nor there. She was managing several interactions at once.

In other situations, a person using a mobile phone sets up a virtual barrier while engaging in mediated interaction. In this way, use of a mobile phone in public spaces points to a kind of buffer between the user and the other co-present individuals. There is a kind of mutual interaction between the "viewer" and the dispensing of civil inattention, on the one hand, and the co-present telephonist, who perhaps requests the indulgence of civil inattention.

Observation   A woman walked along the platform at an Oslo subway station, texting as she walked. She walked with a slow, slightly stiff-legged gait, with her concentration clearly on her mobile phone. Occasionally she would glance up in order to navigate around people standing on the platform.

The use of a mobile phone in a public space often requires a kind of atomized and individualized focus, at least from the perspective of those who are co-present. In the observation here, the woman's posture and carriage were clearly closed to other potential involvements. She had adopted body language that showed her engrossment in the composition of the text message. This shielded her from the limited involvements available at a subway station.

The pose of using a mobile phone is a request for civil inattention. Indeed, the ringing of the telephone and our disappearance into the sphere of mediated interaction while holding the telephone up to one ear can be used to generally remove us from the demands of the local scene.[6] There may even be the sense that we temporarily colonize a portion of the public sphere when talking on a mobile phone. However, this can be an ongoing negotiation with others who are present and who also have a legitimate reason for using the space.

Observation   A man in a bookstore was talking on his mobile phone while browsing through books. It seems that he was getting recommendations or instructions as to which book to purchase (perhaps it was a gift for a third person). He removed a book from the shelf, looked at it, and continued his conversation. I moved into the same section of the bookstore and began to look at books in the same shelf area, about a meter away from the man. I was within his sonic sphere. From snatches of conversation, it seemed he was asking whether a certain book would make an appropriate gift. He moved away from the shelf, yielding to my "interest" in that portion of the store. The man took the book and began to wander slowly in the direction of a cash register that was about 3–4 meters away from the bookshelf. It seemed, however, that he planned to complete his telephone conversation before paying for the book. He sauntered beyond the cash register into a little trafficked alcove/display area while continuing his conversation. After a few moments, I too slowly entered the alcove area and began to look at some items there, never getting closer to the man than about 2 meters. After a few moments he again yielded the space to me and walked unhurriedly out of the alcove area, past the cash register, and back in the direction of the original bookshelf, but he stopped a few meters past the cash register. There were a few other customers in this area of the store, but there was no real press for space. I paused a few moments in the alcove area, wandered out of the alcove area past the man again, and began to look at a book. At this point the man stood between the cash register and me, but closer to me. One last time, he gave way to my interest in that area. He turned and moved slowly toward the cash register and started the goodbye sequence. He concluded the closing, put the phone in his pocket, and began to pay for the book.

Aside from the slightly entertaining spectacle of pushing the poor man around the bookstore, such "gonzo" research has some interesting aspects. I was not a legitimate accredited member of the man's conversation circle, but I was within its audible range. I also had a legitimate reason to be in the store: I was a customer looking at books. Thus, the man was seemingly simultaneously aware of both the need to attend to his conversation and the fact that he was in a public space, with the exigencies that that presented. The strategy was to seek out the less trafficked areas of the store. I was not necessarily willing to grant him the degree of civil inattention in which he was interested.

By refusing to observe the normal terms of deference I was perhaps being too familiar and intruding too far into his private affairs. He did not actually have a claim on the space, and indeed I had the superior right to browse through the books on display (after all, it was a bookstore). Thus, he adopted a strategy of retreating.

Under the rubric of ritual interaction, the man and I were at cross-purposes. Foreshadowing the discussion in the next chapter, the most deeply embedded interaction was probably the man's conversation with his telephonic partner. He was clearly involved in the ongoing maintenance of his personal life. From the perspective of the observer, the call to his wife—if indeed that was the case—had a form and a content that drew on both the broader social patterns for marital interactions and the more specific routines that the couple had developed. The greeting sequences, the presentation of the issue at hand (for example: "Hi, I am at the bookstore. Do you think that Lars would like the new book by Dag Solstad?"), and the discussion around these issues was a reenactment of the couple's sense of identity. There was the assembly of the group—in this case a dyad—with at least some barriers to outsiders. There was a mutual focus of attention and a shared mood. There was also some form of collective entrainment that "played on" and "played into" the solidarity of the couple. In some small way, the interaction may have contributed to the maintenance of their coupled identity, or, to put it in the terms developed by Collins (2004a, p. 48), the interaction potentially revitalized the symbolic validity of their coupled identity, an identity—if my guess about the situation is correct—that is partially founded on purchasing and proffering of gifts to friends and relatives for birthdays, weddings, etc. This was in all likelihood a minor event in their lives. The man, for example, did not appear to be in the middle of an existential argument with his interlocutor. It was not a huge epiphany but rather a simple reconfirmation—another minor episode in their common history.

In addition to the interaction with his telephonic partner, the man had to negotiate the rituals of being a customer in a store. In crowded portions of a store, we may draw our arms in as we pass near other customers, shuffle sideways as we move through tight passages, or stop and start as we take the movements of others into account. That is, we adopt the comportment of being a customer in a store and recognize the needs of other customers. In this case, the store was not crowded, but he still needed to take other shoppers into account. He had to retain a certain kind of demeanor and, in the context of using the mobile phone in a public place, negotiate his implicit request for privacy with his conversation partner with respect to the movements of the other customers. That is, his use of the mobile phone made an implicit request for civil inattention. In terms of Collinsian rituals, while the level of entrainment with fellow shoppers and employees was far below the level of entrainment with his telephonic interlocutor, there were nonetheless issues to which he and his fellow shoppers had to attend. The willingness to pay heed to these issues is worked out in the common deference and demeanor of the persons in such a situation.

Another issue arises with the development of "hands-free" devices for mobile telephones. Such devices can be wired to a mobile phone's handset, in which case they look quite similar to the earphones of a Walkman or an iPod. In other cases, they can be wireless and somewhat more massive, perhaps looking like an oversize hearing aid. In either case, but particularly the former, the mobile telephone is not necessarily in sight, and thus there are fewer visual cues for other co-present persons. They allow the mobile phone user to carry on a conversation via mobile phone wherein the phone may be nowhere in sight. The speaker need not assume the traditional pose, with a handset held to one ear.

There are several issues here. First, there is the sense that the person speaking on the telephone is not providing the others who are co-present with the appropriate visual and audio cues. Because of this, the actions of the telephonist are not accountable. There can be the impression of an individual who is seemingly "talking to the trees." While this may be an interesting exercise, it is not a generally seen as a responsible way of behaving in public situations. It is a breach with the normal sense of social gatherings. A person talking to no apparent other individual violates the sense of how we should present ourselves in co-present situations. It has become somewhat acceptable to speak into a telephone when in public, using the gesture of holding the device to our ear to indicate that we are "on the phone." This is an acceptable—or at least explainable—posture. A somewhat modified version of this is the posture we strike when writing a text message. Again, this implies that we are not necessarily available for interaction since our thoughts are focused on the composition of the message. The point is that others in that social situation can account for our actions. We are doing something that is recognizable and thus we merit the temporary indulgence of the others' civil inattention while we carry out our telephonic business. Hands-free devices, however, present a new challenge.

If we take this one step further, it is possible that when we are using a hands-free device others who are co-present continue to interpret our behavior as though nothing was amiss. This is seen in the example offered by the psychologist Steve Love (2005, p. 1):

A few years ago I was on a train traveling back to Glasgow from Edinburgh. Not long after we started the journey, the girl I was standing next to began to talk to me. I was a bit surprised as I did not know the girl and instead of asking me something like "how long does this train take to get to Glasgow" she asked me how I was and what my day had been like. Although surprised, I told her I was OK; had a good day at work and that I was looking forward to watching the Scotland soccer match on TV later that evening.

As I spoke, I saw the girl starting to glare at me and as she turned away from me I heard her say, "OK, I'll call you later." It was at this point that I realized that the girl had been talking into her hands-free mobile phone microphone and not to me. The rest of the journey continued in silence.

There were two parallel understandings of the situation. On the one hand, like the fellow in the bookstore, the woman assumed that she was in a personal interaction with a close friend. Further, she assumed that the others in the co-present situation would grant her civil inattention. Love, on the other hand, found himself within the personal sonic space of another passenger (Hall 1973). He gamely tried to fit the conversation into the genre of "talk with strangers on the train." Indeed, all the visual and audible cues were in place and only the theme was somewhat awry. Seen from the perspective of ritual, the woman was engrossed in one interaction while Love was considering the degree to which he should become engaged in his interpretation of the situation. However, they were in two parallel universes. The interaction between the two was such an infringement of common interpretation that it led to the negation of further possible interaction. The woman may have interpreted Love's talk as a kind of pick-up ploy, or she may have correctly, and embarrassingly, understood the misinterpretation of her personal telephone conversation. It seems that she felt herself the violated partner and thus the initiative for an eventual rapprochement lay with her. The violation had, however, been too great in her estimation, and so silence was the best strategy.

Mediated Ritual Interaction in Co-Present Situations

In the examples cited above, the persons were aware of their physical location and were also involved in mediated interaction. The exigencies of the mediated interaction influenced their behavior in the local situation. For instance, the main area of interest for the man in the bookstore was his telephonic interaction. Other observations also point to the importance of mediated rituals at the expense of the co-present ones. Though we often pay heed to the co-present, the material indicates that the mediated interaction has an equal if not superior place in the minds of the individuals.

Observation   A casually dressed woman walked out of the area around Oslo's Nationalteater station. She walked east on Karl Johan Avenue,[7] which was being renovated and was closed to traffic. She walked on the southern sidewalk somewhat slowly, trying two or three times to make a call on her mobile phone as she walked. None of the calls went through, and so she began to text. Through the rest of the observation, her focus was mostly on composing a text message. She continued east, walking near the edge of the curb. A woman on crutches walking with two other women approached her, going west. The woman on crutches was also quite near the curb and was somewhat hemmed in by her two friends, who were walking to her left. As the woman who was texting and the woman on crutches approached one another, the "texting" woman gave a "navigation glance" and edged gently to her left. The line of women including the woman on crutches going west also edged slightly to their left. The texting woman and the woman on crutches passed close to each other without any complications. The texting woman then crossed the street at an angle, continuing eastward, and eventually came to the sidewalk on the north side of the street. All the while she continued to text, with her telephone in her right hand. On the northern sidewalk, a portion of the area had been closed off to allow for construction, leaving a passage on the actual sidewalk that was only about a meter and a half wide. At the entry to this "passage," a woman had strategically placed herself in order to distribute handbills. The texting woman approached the handbill woman. The handbill woman took half a step toward the texting woman and offered her a handbill.

The texting woman continued to text, using her right hand. Noting the offer from the handbill woman, she raised her left hand to about waist level with her palm down in a gesture of refusal. The woman offering the handbill retracted the piece of paper and took half a step back.

The texting woman continued into the narrow passage. In the passage, she did not concentrate as much on texting and looked around herself more. After coming back onto the broader portion of the sidewalk she returned the majority of her concentration to texting looking up for a navigation glance only about every 15–20 steps (more often when passing near other people or when crossing a side street.)[8] Eventually she came into the area of Egertorget (a slightly broader section of Karl Johan Avenue that is exclusively a pedestrian area). She continued east, texting all the while. A man who had finished using a mini-bank turned and somewhat quickly walked south while examining the receipt from the mini-bank. The mini-bank man and the texting woman approached each other. At the last moment, he changed his bearing slightly and passed in front of the texting woman. As the man passed in front of her, the texting woman paused slightly in mid-stride and raised her right hand— the one holding the mobile phone—in order to avoid having the man bump into it. Thus, the two came well within each other's intimate sphere momentarily but were able to pass one another without colliding. About halfway through Egertorget, the woman saw three or four friends. (It became obvious that she was en route to meeting this group.) She looked at a male friend and smiled, glanced back to her mobile phone. The male friend took a few steps in her direction and she continued to approach the man. As they came close to each other she used her right arm (the one holding the mobile phone) to give the man a short embrace, unwound her arm from around his neck, looked at her mobile phone again for a short moment to complete her texting, folded the telephone and put it into her pocket, and continued to greet her other friends.

The woman here had to balance between the attention to her texting and the minimalist awareness toward her co-present situation. She showed a certain mastery of the situation by maneuvering past the woman on crutches as she simultaneously composed her text message. She also declined the offer made by the handbill lady, evaded a collision with the fellow coming from the mini-bank, and exchanged greetings with her friend. We see her interweaving of the two spheres perhaps most tellingly in the transition from walking and texting to when she greeted her friends. Indeed her texting was intermingled with the hug she gave to her friend at the end of the observation. She was adroit in her balancing between the textual interaction and the co-present worlds. Indeed, it is not often that we see people moving about the city with their attention so clearly focused on a hand-held artifact. A person who, for example, walks and reads a book simultaneously is often looked askance at. Either it is a hopeless scholar who is not able to tear himself away from some dusty Greek translation or it is a tourist who needs a guidebook for orientation. Thus, the woman's absorption in the mobile phone is an example of a new element in the cityscape. It is an element that is becoming common and indeed accepted. It is important to note, however, that it is not abstract and ego-centered absorption in another thought world; it is social intercourse. It is a conduit through which we maintain social interaction, and we have to pay it heed. Even though we are physically removed from our textual interlocutors, we still have to pay them the proper notice. If, for example, the woman described above had not responded to the text message she had received, it may have been a slight against her texting partner. Just as she was able maneuver physically with those persons around her (the woman on crutches, the handbill lady, the man coming from the mini-bank, and finally the friend she embraced), she had to maneuver socially with her remote partner. That is, she had the responsibility for dealing with two lines of action. In Goffmanian terms, she was acting on two front stages (Ling 1997; Goffman 1959) and in each of these there was a kind of entrainment—however fleeting in the case of the co-present interaction and however phased and interlaced in the case of text messaging—and thus we can see both performances as ritual interaction.

Obviously, the interaction with the texting partner was not available for direct observation. However, the texting woman's interaction with the people she met on the street was there for all to see. It was fascinating to watch her subtle maneuvering around the crutch lady and the others. In these interactions, the different partners drew on a repertoire of gestures. These are devices that we have available as we move about in crowded areas. We give each other signals as we maneuver ourselves along the sidewalks and through the different situations.[9] All of this is a part of our cultural ballast. Through socialization, we have been taught these techniques, and we have developed and reenergized them through countless situations. The observation indicates that the texting woman was sufficiently engaged in the physical setting to maneuver successfully through it. At the same time, she afforded a high degree of attention to the composition and editing of her text message. The engrossment in that activity indicated her engagement in another form of social interaction. In both realms she was observing the ceremonial aspects in their negotiation of everyday life (Goffman 1967, pp. 55–56). The degree to which this was a successful performance must be left to her interlocutor and those with whom she shared the sidewalk. From the observations of the people with whom she came in close physical contact, there was no immediate problem. This is perhaps seen best at the end of the observation. Like the woman observed outside the London tube station, the texting woman adroitly carried off a greeting while still partially engaged in the end of her texting session. The lenience of her co-present friend allowed her to end the texting session and to turn her attention to him. Had she continued with the texting, the physical meeting might have lost its steam and been seen as a failed interaction. There was the danger that, in the words of Goffman (1963a, p. 38), her "heart might not lie where the social occasion requires it to." On the one hand, as she walked up the street she was in a kind of asynchronous textual interaction with an unseen, physically remote interlocutor.[11] As I will explore in the next chapter, the textual interaction with another person is imbued with a variety of the linguistic and social conventions, such as the use of openings and closings (Hård af Segerstaad 2005; Ling et al. 2005; Ling 2005b).

Conclusion

The mobile telephone seems to have several effects when seen in terms of co-present use. It can be used a kind of secondary engagement, as a status object, or as a repository of information and history. As a physical object, it can influence and perhaps even enhance co-present interaction. But it also can be a barrier to co-present interaction. It can disrupt familiar forms of interaction, draw attention away from the co-present, and interrupt out dealings. The discussion here illustrates how in many situations the device has become an obstacle for the once-familiar flow of events (Fortunati 2002; Puro 2002; Rice and Katz 2003). This theme has been examined from various perspectives (Ling 1997; Ling 2004b; Love 2001; Murtagh 2002; Nordal 2000; Palen et al. 2001). Indeed, work by Banjo et al. (2006) shows how mobile phone use inhibits interaction with co-present others. Campbell (2007) describes cross-cultural aspects of mobile communications in the public sphere. This research indicates that interaction via mobile phone takes precedence over co-present interaction, particularly when the interaction is between persons who are not particularly familiar with each other. Seen from this perspective, the device may be disruptive to these everyday rituals. It can cause interruptions to meetings, infringe on our greetings, or affect how we maneuver up and down the street. We are starting to develop a repertoire of gestures that indicate our engagement in other affairs. The "hand to the ear gesture" and the "texting pose" are examples of this. The major point here is that the mobile phone sometimes trumps co-present ritual. It can disrupt the flow of the interaction and direct attention elsewhere. In this phase of analysis, the mobile phone seems to be taking our attention away from those who are co-located, and perhaps it is hampering the ability to generate a common sense of the mood in these co-located settings. In the following chapters, I will examine the opposite: the ability of the mobile phone to draw groups together.

8

Mobile Telephony and Mediated Ritual Interaction

Rituals are not only enacted locally; they are also mediated. Cohen et al. (2007) describe a bris (circumcision ceremony) that took place in Israel. A bris takes place on the eighth day after a male child is born. In this case, the boy's father, serving reserve duty in the Israeli army, was unable to attend in person. However, he was able to participate via a mobile video connection. The people on one end of the transmission were participating in a traditional bris. On the other end, the father was able to remotely play his part. Indeed, the father had an active role. He was not simply a viewer. Thus, there was a ritual taking place in two locations, mediated by the mobile video link. Such a situation is not new. Standage (1998, pp. 128–129) discusses marriage ceremonies that were carried out by telegraph in the 1800s.[1] Although such situations are not common, they indicate that distance does not necessarily get in the way of a good ritual.

In the previous chapter I focused on how mobile communication influences and often disrupts co-present situations. In this chapter, I will focus more directly on the way that ritual is played out via in the actual mobile communication events. In this respect, I am taking a step away from Collins's assertion that co-presence is an essential element in ritual. The main theme here is that groups are able to engender cohesion with the aid of mobile communication. The mobile phone can be used to destroy cohesion or to build it.[2] However, as we will see in the following chapter, the balance seems to be tipping in the direction of mobile communication's supporting the development of cohesion in small groups.

Mediated Ritual

For Durkheim, as I noted above, ritual was, per definition, a co-present activity. Durkheim worked in an era where interpersonal interaction was only rarely mediated. In addition, the objects of his study—Australian Aborigines—may never have come into contact with any form of mediation technology. Goffman worked in an era and in locations with greater access to telephony, but he was for the most part interested in co-present "face engagements" while still recognizing that mediated interaction is an arena in which interaction can take place (1963a, p. 89, n. 12). Collins (2004a, p. 23) asserts that ritual interaction can be developed only in co-present situations, and (with some justification) that it is not possible to develop the sense of mood and effervescence needed to carry off a successful ritual via a remote hookup. Collins discusses, for example, the hypothetical idea of a wedding conducted over a remote link. He notes that the inability of the participants to provide direct feedback would hinder the dynamics of the event. Indeed, events such as weddings (or the bris described above) are rooted in co-present tradition, and it is in that realm that they have their existence (Rice 1987).

It is clear that co-present rituals are powerful motors for the development and maintenance of social cohesion. To quibble with this is pointless. In the broad sweep of social interaction, it is as Collins suggests. Weddings, sporting events, funerals, parties, meetings, concerts, and other events we attend in person are the situations in which we are able to best forge social links. With marginal exceptions, all other forms of mediated interaction pale in comparison to the power of co-present interaction. That said, mediated interaction is also a form of contact through which social bonds can be nurtured. Mediated interaction can influence the staging of broader co-present events. For example, it can be used by operatives arranging a political rally to check on details, orchestrate the timing of the event, and fine-tune the use of effects during the actual rally. Participants can use mediated interaction to anticipate the event and also to recap it afterwards (Ito 2005b). Mobile communication can also be used by participants during a concert, a rally, or a service. For example, participants can broadcast the excitement of being at a concert to others who are not there by means of a "cellcert" (Watkins 2005).[3] Mediated interaction can take its point of departure from the symbols and routines used in the co-present ritual. These can be developed, reinterpreted, or revitalized in the subsequent mediated interaction. The catch phrase from the theater piece or the sermon can be restated and reinterpreted in the phone calls afterward, or the big play in the football game can be dissected in the text messages exchanged between the fans.

Apart from large-scale "Durkheimian" rituals, mediated communication can influence the more Goffmanian forms of interpersonal interaction. As will be described below, the device can be used to facilitate and augment smaller-scale interpersonal interactions. Finally, various forms of ritual can be developed entirely within the realm of mediated interaction. Ways of greeting, forms of argot, and even phrasing can be characteristic of a mediation form. In addition to the communication of information, this metalanguage can serve to underscore that the communication is taking place via a particular kind of mediation and that that in itself is a sign of inclusion.

While co-presence is a major venue for the development of social cohesion, it is also possible to assert that some of the same can be achieved via mediated forms of interaction. Mediated interaction can enhance the broader co-present forms of interaction and can also function in its own right as a means through which members of a group can engage one another and develop a common sense of identity. Indeed, as will be discussed in the next chapter, the directness and ubiquity of the channel can lead to the tightening of social bonds within a group.

These notions will be examined using several different examples of mediated interaction via mobile phone, including greeting sequences, romantic interaction, texting, jokes, repartee, and gossip.

Forms of Mediated Ritual Interaction

Pre-Configured Greetings

Greetings may be the most thoroughly examined of the various mediated ritual forms. Goffman examined them (1967), as did Sacks, Shegloff, and Jefferson (Schegloff and Sacks 1973; Sacks et al. 1974; Schegloff et al. 1977). Goffman suggests that greetings are a way to assure the participants that the "relationship is still what it was at the termination of the previous coparticipation" and that greetings "serve to clarify and fix the roles that the participants will take during the occasion of talk and the commit participants to these roles" (1967, p. 41, n. 30).

Telephonic greetings—for example, "hello" (Martin 1991, pp. 155–163; Bakke 1996; Marvin 1988; Fischer 1992, pp. 70–71)—may be the most ritualized portion of the common telephonic situation.[4] But there were difficulties lurking behind what we now see as a simple greeting. In early telephone conversations in the United Kingdom, there was a need to work out the status differences of the speakers that previously had been handled visually (Fischer 1992, p. 70). It was seen as an affront to an upper-class person to be addressed by a lower-class person with a familiar "hello." Thus, elaborate forms of address were developed, such as "Mr. Wood of Curtis and Sons wishes to talk with Mr. White" or "Hello, this is the Jones residence, Samuel speaking."[5] Time has not been kind to such overblown expressions. By the time Sacks, Schegloff, and Jefferson (1973, 1974) examined greetings, the following was typical: "Hello." "Hi, this is John." "Hi John, how are you?"[6]

Laurier (2001) has discussed how greetings have changed to include geographic location. Beyond geographical information, the mobile telephone has also changed greeting rituals by doing away with the need for the interlocutors to identify themselves.

Observation   A man was standing in a store on a Saturday morning. He was examining some items in the store when his phone rang. He pulled the phone out of his right jacket pocket and looked at it long enough to activate it and read the caller ID in order to know who was calling. Then he raised the phone to his ear and, instead of the traditional greeting sequence, said "Are you ready?" (pause) "OK." There followed a short conversation with two or three turns. The conversation was quickly over and the man replaced the phone in his pocket.

The lack of ceremony in the opening bespoke a call between two persons who were in frequent contact. This is also a good example of pre-configuring the greeting sequence. There was no need for introductions, formal interactions, or the so-called double introduction (Schegloff and Sacks 1973; Sacks et al. 1974). The mobile phone reduces or even eliminates the need for any form of opening, since the person making the call is calling to a specific individual, and the person receiving the call usually has the more commonly used numbers in his or her name register, meaning that the name of the caller appears via caller ID. Since both partners are clued into this, it is perhaps seen as artificial to go through the sequence of

"Hello, this is Rich calling; may I please speak to John?" This is particularly true if the interlocutors are in a connected state. It is, for example, possible to speculate that the man in the observation shown above was only doing a short errand while waiting for his wife to finish another task. The call was meant simply to alert the man to his wife's current status with respect to the broader project of shopping on Saturday. They were so engrained in each other's situation and so aware of the other's general context that the abbreviated messages were easy to unpack (Miyata 2006).[7]

If the telephonic interlocutors are not intimate, a more extended greeting might be called for—one that allows the partners to rearrange their mental furniture in order to place themselves into the context of the call: "Oh, is that you, Sally? It is good to hear from you. Is Dick doing OK? What about Spot and Puff?" If this were the case, then the two interlocutors would have to go through a process of establishing the context of the call and deal with the issues outlined by Goffman. They would need to remind themselves as to the nature of the interaction with the somewhat more distant person. In the case of the call reported in the observation, there was no need for that, since the call was a kind of "connected presence" (Licoppe 2004). Speaking of textual interaction, Ito and Okabe (2005) describe how a mobile phone can be used to give a "virtual tap on the shoulder":

These messages define a social setting that is substantially different from direct interpersonal interaction characteristic of a voice call, text chat, or face-to-face one-on-one interaction. These messages are predicated on the sense of *ambient accessibility*, a shared virtual space that is generally available between a few friends or with a loved one. They do not require a deliberate opening of a channel of communication but are based on the expectation that one is in "earshot." . . . As a technosocial system . . . people experience a sense of persistent social space constituted through the periodic exchange of text messages. These messages also define a space of peripheral background awareness that is midway between direct interaction and noninteraction. (Ito and Okabe 2005, p. 264; emphasis added)

Although Ito and Okabe are describing interaction via text, many of the same points apply to the man described in the observation. If I interpret the situation correctly, there was a kind of ambient accessibility between the two, and there was a general background awareness that the call would come at some point within a particular time frame.

It is also instructive to consider the loose greeting in the observation. As a child, in the age previous to mobile telephony, the man probably was instructed on how to answer the phone using a variation of the more formal introduction where greetings and names are exchanged. However, since the advent of caller ID and the direct interpersonal calling implied by mobile telephony, the more decorative aspects of the greeting have obviously been jettisoned.[8] Had the man used a full formal greeting sequence (including phrases such as "This is John Smith" and "To whom do I have the pleasure of speaking?"), it would been seen as strange. Just as we perhaps greet work colleagues once in the morning, but then do not need to repeat the greeting through the day as we meet them in the hallway or in the lunch room, the man in the store described here obviously did not feel the need to re-greet his interlocutor. Thus, while the initiation of the phone call may have seemed abrupt to other's ears, to behave otherwise may not have been appropriate in the context of that dyad.

The greeting sequence in the observation reported above was certainly a more streamlined version of the traditional greeting sequence. But if I read the situation correctly, the greeting that was proffered, however perfunctorily, may have been the most correct one. It gave evidence to the person calling that the man who took the phone was paying attention and providing the appropriate amount of focus. The man in the store gauged the mood of the interaction—slack time while out shopping on a Saturday—and paid it the proper heed. In addition, the interaction allowed for the further interaction between the two persons.

In some respects, the interaction in the observation is boring. Not much happened. A middle-aged man received a call on his mobile phone, grunted a couple of utterances before hanging up, and continued on his way. But in fact some important things happened. The observation seems to indicate that he maintained the flow in an ongoing interaction with a well-calibrated style, using the appropriate symbolic devices. Had he been much more or much less enthusiastic, he would have sabotaged the goal of the telephone call. As it was, he managed to account for himself and his actions. Was this a ritual interaction? If we tone down the definition of a ritual interaction, then we can assert that there was mutual engagement in the continuing interaction between the man and his interlocutor. Thus, though it was not done with the accompaniment of drums or with intense Durkheimian effervescence, the interaction fulfilled Goffman's notion of mundane yet ritualized activities.

Using a slightly different test as to whether this was a ritual, we can speculate as to whether the interaction could have failed. Again the answer is Yes. The man could have ignored the call; he could have taken the call but not said anything; he could have flown into a rage. In each case, the interaction would have failed as a ritual. As it was, however, he carried it off.

Negotiation of Romantic Involvements

Mobile communication is used in the cause of romance to support co-present interaction. It allows people to exchange comments and endearments when they are not physically together.

It is clear that the mobile phone has affected romantic involvements. In an oddly unbalanced finding, material gathered in Norway shows that 50 percent of teen girls and 32 percent of teen boys flirted by mobile phone at least once a week.[9]

Within a closed group, romance is one of the more ritualized forms of interaction.[10] Indeed, the negotiation of a romantic interest is perhaps the quintessential small-scale ritual. When two people are entering into a relationship, there are hints, glances, and staged interaction. Nearly every word or gesture is interpreted. Various techniques and devices are drawn upon in this small-scale Goffmanian ritual—for example, wearing special perfume, smiling and laughing at comments, the offering of a rose, the planning an intimate meal.[11] If nuanced interaction does not carry the day, a broad hint or a risqué double entendre may. After the courting has begun in earnest, there are various expectations placed on the partners as to their mutual availability, the forms of attentions, and the kind of interaction that is to be expected.

Collins (2004a, p. 230) discusses flirting as an interaction ritual chain. He discusses the initiation of flirting via, for example, touching feet underneath the dinner table (ibid., p. 240). Goffman (1971) discusses how the "tie sign" of holding hands helps to define the couple for each other and in the public eye. It is beyond question that co-presence is, in the vast majority of cases, an essential element in the development of a romantic relationship.[12]

As Collins notes, the assembly of the group (in this case often a dyad), the mutual focus of attention, the establishment of a mood, and the barrier to outsiders are all elements that are necessary for the establishment of coupled solidarity. While physical co-presence is often a key element at various points in the development of a romantic relationship, mediated interaction has also been a discreet aspect of what he calls "linking" in the courting process (Collins 2004a, p. 193). Messages and "back-channel" communication have often played a part; indeed these messages are often the handmaiden of love. It is in this way that important information on the other's availability and one's own intentions are made explicit. In literary history, Romeo's ill-timed notes to Juliet and the various missives between Darcy and Elizabeth in Jane Austen's *Pride and Prejudice* illustrate this form. In my own past, passing notes in class (folded in a particular origami-like way) was how nascent couples started a relationship. Thus, there has long been the use of mediation technology in these enterprises. The telegraph (Standage 1998, pp. 134–136), the traditional telephone (Fischer 1992, p. 234; Marvin 1988, pp. 67–68), and now the mobile phone (Andersen 2006; Byrne and Findlay 2004; Ellwood-Clayton 2003; Fortunati and Manganelli 2002; Habuchi 2005; Kim 2002; Ling 2004b; Prøitz 2006; Solis 2007) have been used in the exchange negotiation of "love projects" (Prøitz 2006).

The initial stages of a relationship are always troubling, particularly for teens. Determining the other's suitability and level of interest and the kind, amount, and tempo of self-disclosure that is appropriate are all problematic. The way with which to deal with these issues has changed. The cycles of being in the public square at the appropriate time and promenading in the accepted places was replaced by interaction by telephone. This again has changed with the adoption of the mobile phone. Material from group interviews in Norway reveals that after meeting a possible love interest, the hopeful couple exchange mobile phone numbers and start a more or less extended and more or less explicit series of text messages (Ling and Helmersen 2000; Ling 2000; Andersen 2006). Rather than synchronous interaction by telephone, in which it is perhaps easy to make a false step, the couple engage in the exchange of carefully edited text messages. As they gain their footing, the frequency of the messages might increase. This indirect form of interaction allows the individuals to cover over some of their more obvious character flaws and to move through the preliminary stages of a relationship in a more deliberate way. It also is a way of saving face in the case of giving or receiving rejection.

Rita, 18: . . . if you meet a guy when you are out for example, then it is a lot easier to send a message instead of talking like. Somebody you don't really know. It is more relaxed.

Anne, 15: It is easier to tell if you like a person.

Interviewer: Via SMS?

Ida, 18:   Then your voice will not either shout or disappear. You need time to think [when constructing your messages].

SMS also makes it possible to contact the potential partner without the fear of being directly rejected, something that is more difficult to manage in voice-based interaction. Andersen (2006, p. 45) reports this exchange:

Espen: It is super dumb if you call and you hear that the hag is really uninterested, that she just doesn't say anything.

Cato: Embarrassing when it is completely quiet . . .

Espen: "Why are you calling me?"—you know, that is not cool.

Further, it is important that the message or messages used when flirting have a kind of deniability. This is afforded by playing on humor. Teens note that they send messages that are on the "fresh" side, such as this one reported by Andersen (ibid., p. 53):

Im an alian and I've transformed myself in to your phone. While you're reading this Im having sex with your finger. I know you like it because you're smiling;)

If the appeal does not find approval, the author can simply say that he was just kidding and avoid loss of face.

The timing of the message receipt/response cycle is also important. In parallel with how many times a phone is allowed to ring, it is possible to respond to a text message too quickly or too slowly. If the game is played successfully, the two individuals are drawn into the same humor with regards the potential for the relationship. As this happens, the frequency of messages might increase and they might move into more synchronous forms of interaction, such as PC-based instant messaging, voice calls, and face-to-face interaction.[13] The voice call has obvious advantages (and dangers) in terms of developing mutual engagement. More tightly interwoven turn taking, mutual rhythm, use of tonal range, and use of pitch help a couple develop cohesion (Collins 2004a, p. 48). If these elements are handled well, they can further encourage the relationship. If one partner dominates the conversation, or if one partner is halting or awkward (in other words, if the couple experiences a failed ritual), it is likely to be a damper on the development of the relationship (Andersen 2006). The effect of this process is to establish the mutual recognition of a common mood within the dyad. While the co-present interaction is a key to the establishment of a relationship, the mediation of information via other channels can also influence the establishment of a more serious relationship.

■

Once a relationship is established, mobile communication influences how the partners keep each other appraised of their status. Material from Norway shows that there is the exchange of endearments by lovers in the form of "good night" and "good morning" text messages. Examples range from the utilitarian "G'nite" (female, 15) or "Have a good night, hug" (male, 47) to the rakish "Do you want to spend the night? Hug" (female, 17) and "Good night sex bomb" (female, 35). Others are more mysterious: "Hug, has the prince slept well? Use the signal" (female, 48). Some are openly infatuated: "Nite, I love you" (female, 27). "I wish you a good night and I love you" (female, 28). Others refer to the internal lingo of the relationship: "Good night and sleep with an image of a sleeping bear" (female, 16).[14] In the case of teens who still live with their parents, this represents a private communication channel where others are not available. This form of interaction is ignored only at great risk. Ito and Okabe (2005, p. 265) report many of the same types of interaction among Japanese teens. They note that for couples living apart "messaging became a means for experiencing a sense of private contact and co-presence with a loved one even in the face of parental regulatory efforts and their inability to share any private physical space." Consider these messages, reported by Ito and Okabe:

I really want to see you (>_<). I am starting to feel bad again. My neck hurts and I feel like I am going to be sick (; _;). Urg.

I get to see you tomorrow so I guess I just have to hang in there! (^o^).

These messages (by a young Japanese couple) include both text describing the emotional state of the individual and the characteristic Japanese emoticons. These comments are included in a longer, chant-like sequence of remarks. One partner was using the public transport system and the other was presumably at home. The ability to tuck these interactions into otherwise empty time is a unique characteristic of mobile communication. In this way the relationships becomes more omni-temporal. Texting allows the couple to maintain discreet, continual, and in some cases intense contact with one another. The interaction stretches from when the two part at school through the evening (including time spent doing homework, watching TV, and so on), interrupted only by bathing and sleep (ibid., p. 138). Indeed, teens in Japan, and also in Norway, have commented on the need to manage the communication flow so as to allow themselves time for private activities. Thus, although texting or using another social networking system is not a co-present activity, the partners are intensely focused and a common frame of mind is shared.[15]

In language that approaches the notion of ritual induced cohesion, Ito (2004) comments that "this steady stream of text exchange, punctuated by voice calls and face-to-face meetings, defines a kind of 'tele-nesting' practice that young people engage in, where the personal medium of the mobile phone becomes the glue for cementing a space of shared intimacy." In the next chapter, I will examine this in terms of Licoppe's notion of connected presence.

■

In addition to facilitating remote interaction between partners, mobile communication helps to enhance co-present interaction. Ito and Okabe (2005) call this an "augmented flesh meet." They describe how two young people in Tokyo start the process of having a date by exchanging text messages during the hours before their actual meeting as they complete their work or their studies and as they negotiate the public transit system. After the date, they return to textual interaction as they go their separate ways via the Tokyo transport system and dwell on "the fading embers of conversation and contact." It is easy to see that the richest time is during the co-present phase of the date. However, the coming interaction can be planned and negotiated beforehand. Afterward, via mediated interaction, it can also be relived and reified as a part of their couple's lore (Ito and Okabe 2005, p. 271; see also Andersen 2006, p. 55). Were the battery to give out or were the connection to be lost because of spotty coverage, the interaction would be considered a failure.

A similar profile is seen in the exchange of text messages leading up to an hour-long telephone call on an evening when they will not see each other. This is then followed by a series of text messages in what Ito calls the "afterglow" of the telephone conversation. These text messages comment on the telephone conversation and perhaps embroider the various themes that were discussed in the telephone conversation. The teens use text messages to extend the face-to-face meeting or the telephone conversation. This does not make the co-present time less important; rather it may respectively foreshadow it and give them time to reflect on it from a slight remove. Like a cubist painting, it may expose the same event to multiple anticipations and interpretations.

The ability to extend the interaction beyond the period of co-presence (the actual point at which an individual "shows up") is also somewhat more diffuse. So long as there is interaction (mediated or co-present), the individuals are at least sharing time (Ito and Okabe 2005a). It is true that the real business of being a couple is best achieved when the two are together, but the preparation for the time together and the reflection on recently shared time are extended and blurred.

▪

My discussion up to this point paints a rather puritanical picture of dating. There are obviously other more carnal forces afoot, and these are also being played out via the mobile phone. In research carried out in Norway, for example, there is a covariance between mobile phone use and sexual activity (Pedersen and Samuelsen 2003; Ling 2005).

The Norwegian researcher Lin Prøitz has assembled a corpus of text messages that go beyond the rather coy discussions of flirting and finding partners into much more explicit territory. In this material, there is the expression of much clarity with regards a range of sensual desires (Prøitz 2006). An example of this is the text message from an 18-year-old girl: "It's insane . . . Just met you three times, but have never felt anything like this before . . . Want so much to hold you . . . Kiss you . . . Going crazy!" In Prøitz's messages we can read about teens who express the unbearable longing to be with paramours—if they can only separate themselves from current boyfriends or girlfriends. They reflect on their own sanity while apart, and they are lavishly frank in their description of their desires.[16]

Bella Ellwood-Clayton, who studied texting in the Philippines, found many of the same forces at work. Texting has developed as an alternative to the more staid system of formal courtship. Consider this example (Ellwood-Clayton 2003, p. 233):

Leticia: Does it mak u hapy 2 stel a kiss? Remember tho shalt not stel, best to ask!

Captain: I kno wen 2 do it

Leticia: Com show me how so I myt also

Captain: Jaz lyk a magican . . . I nevr reveal a secret . . .

Leticia: (sent a message with the graphic of a dancing bear)

Captain: wers my kiss?

Leticia: Y dnt u cum n get it? Latr, im nt yet going 2 bed. I's stil her at d prayer meetn, prayn 4 you. . .

Captain: Ur not prayn. Ur thnkn of me.[17]

This brings us back to the question of just where the ritual interaction is being carried out. The woman at the prayer meeting would seemingly fit nicely into the Durkheimian notion of focused co-present interaction. She should be drawn into the prayer meeting, sharing sidelong glances with other worshipers and joining in the common chanting and singing. She should be feeling the mutual effervescence of the co-present event. It should be such that the tempo of the sermon and the singing of hymns should draw her, along with the others who are physically present, into the flow of the event. Instead, it seems that the chief occupation of her thoughts and the chief form of cohesion being developed is associated with her eventual meeting with the Captain.

Ellwood-Clayton describes how the texting relationships exist in the gray zone between gabbing, entertainment, and earnest desire for a partner. The text relationships are in some cases only short-term flirtations; in other cases they solidify into actual (read: physical) relationships. There is, of course, the complexity of any affair of the heart where one partner is more serious than that of the other.

The work of Prøitz and Ellwood-Clayton goes beyond the rather more innocent texting of Ito and the material from my analyses. In the context of ritual interaction, the material from Prøitz and Ellwood-Clayton seems to include a lot of what we might consider sexual foreplay, albeit in a virtual mode. There is seemingly little question, however, that by the time that Prøitz's couples met negotiations would be at an advanced stage.

In the rather dry language of ritual interaction, the social boundaries as to the in- and out-groups would have been well established, the effervescence of Durkheim and the common mood and entrainment of Collins would be well in hand, and there would be a high degree of mutual engagement in a small-scale Goffmanian ritual.

Text Messages and the Definition of Group Boundaries

Moving from romantic involvements to the development of texting argot, I want to examine a form of interaction that is characteristic of mobile communication. Where romantic interaction often involved a combination of co-present and mediated interaction, the development and use of texting argot is more firmly based in the realm of mediated interaction. The use of idiomatic formulations is often based on groups that have a co-present arena of interaction. While some of the argot may be shared by the IM world (Ling and Baron 2007), there are features of texting that are unique to mobile communication.[18]

The form and content of texting is an area where the group develops ritual forms of interaction. While this form of interaction is not as co-temporal as other forms of mediated interaction, I will suggest that there is nonetheless the potential to establish the mutual engagement and common mood of the Goffmanian ritual. In particular, when the exchange of text messages is at its most intense, the interaction takes the form of a somewhat halting but pathos-laden conversation, and in this mode it can have ritual dimensions (Döring and Pöschl 2007).[19]

The specific locutions that are used in text messages and the phatic nature of many text messages show how the group uses this form of interaction to mark membership (Williams and Thurlow 2005). The use of slang to define adolescent groups is, however, nothing new (Schwartz and Merten 1967). In regard to co-present language use, Riesman et al. (1950) wrote: "Language . . . becomes a refined and powerful tool of the peer-group. For insiders language becomes a chief key to the currents of taste and mood that are prevalent in this group at any moment. For outsiders, including adult observers, language becomes a mysterious opacity, constantly carrying peer-group messages which are full of precisions that remain untranslatable." This capacity has been carried over to mobile communication and into texting. Just as with fashion in clothes, there is fashion in the argot of texting. This can be seen in, for example, the orthography of the text messages. Among teens in Norway it was common to substitute the letter z for s-sounding letter sequences. This was seen in closings such as 'koz', 'klemz', and 'hugz' (all Norwegian variants of 'hug') (Ling 2004c; Prøitz 2006). It was also seen in the use of ancillary words such as 'lizzom' as a substitute for 'liksom' (literally "like," as in "That was like a real good dinner) and 'azz' as a substitute for 'altså' (meaning something similar to "you know," as in "He is a good looking guy, you know"). The practice even was used in the formation of possessives, such as 'dokkerz' (meaning "yours").[20]

Substitution of z for s became characteristic of young urban middle-class teens (read: teenyboppers). A Norwegian rap band called Karpe Diem lampooned this in their song "Fjern deg" ("Get Lost"). In view of the popularity of the band, this put a stop to the practice, as no one would want to be seen as using a form that was seen as naive. While the practice still filled the function of marking a group boundary, it had become the butt of parody.

The teens, however, were not to be stopped. For whatever reason, some Norwegian teens settled on the use of pidgin Swedish as a marker of in-group communication. They used, for example, the word 'tjenara' (a misspelling of the informal Swedish greeting 'tjenare'). They also used 'krämmar' (a misspelling of the Swedish word 'kramar', meaning "hugs") as a closing. The point here is not to be orthographically correct; the point is to have an argot. Given the loss of the previous kind of slang, a new one that was readily at hand was adopted without missing a beat. The transition between the z formulations and the faux Swedish formulations shows the fluidity and the importance of boundary establishment for these teens. When the one symbol set had outlived its usefulness for the group, it was dropped like last week's laundry and a new idiom was adopted. It is also interesting that it was not simply a single word that had been dropped, but rather a whole style of spelling.

Spelling in text messages has indeed been the subject of much discussion.[21] It is interesting to note that both 'tjenara' and 'krämmar' are more complex and difficult to write than possible alternative words. Moreover, 'tjenara' is obviously much longer than either the Swedish ('hej') or the Norwegian ('hei') equivalent of English 'hi'. 'Krämmar' is also one letter longer than the correct Swedish spelling (kramar), and with its umlaut it requires five additional keystrokes.[22] Thus, the teens are using longer, more complex, and more ritually involved words in their messages at the expense of writing efficiency.[23]

The use of writing devices does not end with the spelling and selection of words. Teens also draw on capitalization and punctuation when examining in-group and out-group connections. Teens comment, for example, on the fact that older people make what is to them an obvious faux pas by writing their messages in all caps (Prøitz 2006).[24] Elderly people's—and here teens usually mean anybody over 30—perceived inability to master the technology is seen as another boundary marker (Ling 2007b; Preece

2004, p. 58). Another area where we can see the assertion of generational boundaries is in the use of punctuation. In particular there has been the analysis of ellipses as a form of punctuation that characterizes both text messages and IM interaction (Ling and Baron 2007). According to teens, this adds an oral dimension to their writing. An ellipsis becomes something like an all-purpose punctuation mark. An ellipsis can be used as a period, particularly when implying a dramatic pause:

I mean, I want to see you . . . I'm just stresssed & overwhelmed[25]
Derek got married last thurs. . . . just found out . . . wtf?

An ellipsis also can be used in place of a question mark:
how come you called me so late last night. . . .

And it can be used in place of a comma:
Sooooo . . . i have been doing a lot of spying lately

When asked why not use a more traditional form of punctuation, one teen replied "That is what my mother would use to write," indicating that even the punctuation has a group definition and a boundary-maintenance function.

∎

It is clear that the above examples in themselves are only one part of a broader project of group identification, or what Fine (1987, p. 128) calls "group culture." It is also clear that these words and punctuation marks will be exchanged with others in the next turn of the teen culture cycle.

However, the use of text-based slang is a part of this group's common argot. Like other aspects of fashion, it is in continual flux. Fashion is a balance between the avant garde and the dowdy and a balance between individual expression and group allegiance (Simmel 1971, pp. 304–305). It is also an indication that the society is dynamic and not moribund. It is through the discussion of fashion, however absurd or drab it may be, that we work out some of our collective sense of social order (Duncan 1970, p. 268). It is, however, in the attribution of symbolic power to such features (e.g., a tattoo, a telephone, an oddly spelled greeting) that the group identifies its center and its circumference:

A group is obviously dependent on physical objects, on inherited systems of signs, and on the biological characteristics of individuals acting within the group, but these do not in themselves constitute the group. Only when signs are invested with meaning and emotion can they develop attitudes. The group as an object of reason, faith or emotion is created and sustained in the communication of attitudes, but these attitudes must be objectified in concrete symbols before we can act together. (ibid., p. 431)

As with the adoption of particular types of mobile telephones, the use of a particular style of language is a part of the group identification process (Ling 2001). This can include rejection of a style when it becomes passé. The mutual use of these terms marks a joint sense of expression. It is the lingo of the group, and the group members use it to mark their common project of sustaining the group. It provides them with a common footing upon which they can build when they meet.

■

Argot is an example of Simmel's (1971, pp. 304–305) "joint spirit" being worked out. Like lovers' exchanges of endearments, these special phrases and words (whatever they are at the moment) are important in this process. To not use them is to is to place yourself above the group. To use them is to mark your membership in the group.

The point is to mark the social boundaries. Clothing and fashion have been traditional ways of marking group and status differences (Flugel 1950; Lynne 2000; Davis 1985; Davis 1992; Tortora and Eubank 1989). From the ranks of the adolescents, we can see the use of slang among jocks, debutantes, skaters, and various gender-based groupings. Simmel (1971, p. 304) writes: "From the fact that fashion as such can never be generally in vogue, the individual derives the satisfaction of knowing that as adopted by him it still represents something special and striking, while at the same time he feels inwardly supported by a set of persons who are striving for the same thing, not as in the case of other social satisfactions, by a set actually doing the same thing."

Fashion graphically shows the individual's group interest or group association. What artifacts we use, and how we display them, can mark our status and our association with a particular group (Lynne 2000). When successful, it results in a social cohesion or "a kind of social euphoria which binds us together simply because we are together" (Duncan 1970, p. 269; see also Katz and Aakhus 2002a; Wei and Lo 2006).

The mobile telephone—that is, the physical object or the handset—is a kind of fashion accessory. Just as the decision to have the collar of your

Lacoste shirt up or down communicates an aspect of social membership, so does your selection of a mobile phone. Teens understand this:

Eskild (13):   It is not fun to have the worst [mobile phone] on the market you know.[26]

Inger (17):   I have a real ugly Bosch telephone . . . it's so big and awkward.

Nina (18):   It is . . . about how it looks and its size. Often it is the small cute mobile telephones that have the most status, at any rate that is how I experience it.

The same things that can be said for the selection of clothing can also be said for the use of slang. The turning of particular phrases and the use of specific words in specific situations are also ways of celebrating membership in a group. Interestingly, the use of z endings or faux Swedish as an argot is found only in electronically mediated interaction, and most particularly in text messages. The teens cannot enunciate the z ending, nor is 'hug' an appropriate verbal closing statement when parting. It is not a part of the broader written culture, nor is it a part of the verbal culture (Baron 1998, 2000). These small markers are, then, parts of the broader work of establishing and working out group solidarity that is used only in texting. They are spawned in the mediated interaction, perhaps commented on and laughed at in co-present meetings, and used in texting by the local cognoscenti. This is one of the small bricks out of which the edifice of group solidarity is built.

This form of interaction also brings up another question, namely co-temporality. When we speak of mutual engagement, this often means co-temporal interaction. In the sense of ritual developed above, there is often the sense of common mood and joint entrainment that implies co-temporality. The establishment of a common sense of cohesion via the use of texting argot indicates, however, that the ritual interaction might have a temporal dimension. On the one hand, the use of argot can remind us of a totem—a symbolically imbued artifact that bears the sense of the group between more intense interactions. At the same time, the more tightly intertwined SMS sessions can approach the form of a conversation. In this mode, the content of the interaction can engender the mutual engagement and common mood of the Goffmanian ritual.

Repartee as Ritual Interaction

Another form of mediated ritual is repartee. Recently I was on a bus and a man sitting in the seat immediately behind me received a call on his mobile phone. He took the call and started to speak in a language I did not understand. The only portions of the conversation I understood were his periodic laughs. He would make an utterance and punctuate it with a laugh. He would make another utterance and than laugh again. He would listen for a moment and then laugh. Then he would make an utterance and laugh again. There were portions of the conversation where there was more talk and less levity, and then he would go back to the utterance/laugh cycle once again. After 5–10 minutes, he went through what must have been a closing routine because he stopped talking, closed his phone, and put it back in his pocket. This was an interesting observation because, as noted, the only intelligible part of the whole interaction to my ears was the laughter that made up a not inconsiderable part of the call. As this incident points out, humor and repartee are part of mobile communication. In this section, I will outline how mediated humor is a ritual form that facilitates in-group cohesion.

In all likelihood, there was a carefully timed interlacing of comments between the fellow I observed and his interlocutor. One person played off the other in a well-choreographed interaction that provided enough space for each and also avoided pauses and overlapping in the interaction. Collins suggests that a central feature of interaction rituals is the adoption of a shared rhythm that results in engagement and, in many ways, pulls the individuals into the flow of the interaction (Collins 2004a, p. 48). To the degree that we achieve this, we will feel that the interaction has been a success. If one or the other interlocutor dominates or is too passive, the interaction will not provide the same incitement toward cohesion.

Learning to finely interlace interaction on the telephone comes with experience. Sharon Veach has examined how, as children mature, there are usually fewer and shorter pauses in their telephonic interactions. Veach (1980, p. 276) also notes that speech overlaps actually increase for the oldest children in her sample. Though we might have expected conversation overlaps to decrease as we gains competence, it is also easy to see that as one gains experience in conversation one may begin to use the various grounding phrases discussed above to indicate engagement in the discussion. Once a child gains a sense of precision with regard to the timing of a telephonic interaction, he or she can begin to use it to express meaning and emotion in interactions, making fuller use of what Collins calls "rhythm." This brings interactions to life. Rather than the largely one-way interactions between, for example, a grandparent and a young grandchild, older children can begin to use the telephone for repartee and engaging social interaction. Thus, rather than indicating a lack of skills, the use of overlapping comments seems to underscore the growing mastery of telephonic interaction. As with the fellow I observed on the bus, it suggests that a child or an adolescent can begin to use telephonically mediated interaction as a link in the development and maintenance of social cohesion.

Humor has been a part of a part of mediated interaction from the time of the letter and the telegraph (Standage 1998, pp. 65, 132). The internet and email made gag emails, emoticons, greetings, and ASCII-based representations possible (Danet 2001). Humor, joking, and repartee engender social cohesion, help in integration and in-group identity, provide insight into the exercise of power relations and repair work within groups, and mark group boundaries.

In a 2002 survey, about one-third of the surveyed users of mobile phones in Norway said that they sent a text message containing a joke at least once a week. There are significant age-related effects. About 54 percent of 13–18-year-olds reported sending jokes weekly, while only 12 percent of 45–54-year-olds and only 4 percent of those over 65 reported the same frequency.[27] It also appears that females are somewhat more likely to send jokes than are males. While about 5.2 percent of males reported sending jokes daily, 8.6 percent of the women reported the same behavior.[28]

In the observations and in the analysis of mobile text messages it was common to find people who used humor in a variety of ways. Jokes, humorous quips, and laughter are common in mobile communication. Here are two examples[29] of jokes sent via SMS and collected from respondents:

If you come home to a man that is warm and loving you have come to the wrong house. (42-year-old female)

A couple takes a drive after having argued. They see some pigs and donkeys. The man asks "Some of your relatives? The woman answers "Yeah, the in-laws." (45-year-old female)

Here, the same text message includes both the setup and the punch line. Such messages do not follow the pattern of a kind of call and response, as in "Have you heard the one about . . . ?"[30] The fact that these were transmitted, however, indicates that there was a jovial frame around the textual interaction of the interlocutors. In addition, if the receiver of the messages thinks that they are humorous, it will reflect on the future inter-action between the sender and the receiver.

Evidence of ongoing text-based patter between individuals can be seen in text messages such as these:

Haha now ur in my boat (19-year-old female)

ha so funny . . . r u done? (20-year-old female)

The messages that stimulated these comments are not a part of the corpus, but it is easy to speculate that these replies are a kind of a "comeback" based on an earlier comment.

These examples of mediated jokes and repartee follow a general outline of a humorous story or interaction. In the case of the first two there is a narration followed by a punch line; in the case of the latter two it is easy to speculate that they are humorous rejoinders to some sort of previous utterance.

Humor, when done skillfully, develops a common focus, engenders a mutually recognized mood, and results in social integration in a small group.[31] In short, these forms of interaction are a ritual. When telling a joke, there is the establishment of the context in which a funny story might be appropriate, the introduction of the tale and a series of more or less familiar elements leading to the punch line. Repartee is somewhat different. Instead of a narrative told by one person, perhaps with scripted participation of the audience, both sides of the conversation have the freedom to provide a barb or a zinger as the interaction develops. One person may play the "straight man" while another plays the foil.

In league with co-present interaction and co-present humor, mediated humor provides for social integration and in-group solidarity. Humor is a way to assert power without the need for coercion, and it allows for various types of social repair work. Finally, humor is a way for a group to mark boundaries.

Humor and Group Solidarity   Jokes and repartee often involve the establishment of an in-group. This is particularly the case when the butt of the joke is another group in society (Brown et al. 1987). When done correctly, there is a mutual focus when telling a joke or engaging in banter, there is entrainment in the situation, and it can help in the work of social cohesion. When we tell a joke and others laugh, we experience Durkheim's mutual effervescence. Indeed, laughter marks the existence of collective effervescence and provides a sense of social euphoria that strengthens and rejuvenates relationships and thus reinforces social bonds (Duncan 1970). Telling or hearing a joke or participating in repartee is an example of a collectively produced event (Collins 2004a, p. 65). All concerned are involved in the same symbolic universe. Duncan (1970, pp. 257–258) writes:

Humor in itself is a kind of bridge, a passage by incongruity from one view to another which society provides as an escape from the crushing weight of traditions or the painful anxiety developed by conflicting loyalties. . . . The forms of passage offered in humor, the many ways that we reverse rank, the inner Saturnalia and Feast of Fools which humor brings to the soul are not simply ways of "relieving tension" but of keeping social bridges in good repair. Without such bridges we become islands and the self is powerless until new ways of reaching real others is established.

Both the person telling the joke and the listeners can imagine the same chicken crossing the road or the same group of people trying to screw in a light bulb. If the joke or banter takes a more cutting direction, the traveling salesman and the farmer's daughter are up to no good or the underachievers (however they are defined) are having a tough time. In the case of banter, when the people involved are on the same wavelength, the bon mot of one person gets played back with an imaginative twist that only heightens the effect.

Co-presence is not a requirement for telling a joke. It is obvious that one can use a mobile phone to engage in light-hearted banter.[32] Laughter and light-hearted references facilitate interaction with intimates. For example, laughter can be used to mark the pre-existing solidarity of a relationship. This can be seen in the following observation:

Observation   A middle-aged man in casual clothes was walking toward the street exit of a department store on a Saturday morning. It was raining outside, so he paused near a display window while he dialed his mobile phone. After dialing, he raised his phone to his ear and waited for the party on the other end to take the phone. The other party answered and the man's portion of the dialogue was approximately as follows:

Hi, I was in Pentik's (a local store) and they did not have it (laugh).
(response from other person)

Yeah, I don't know.
(response form other person)

Yeah.
(response from other person)

OK, yeah bye.

The man then glanced at his phone, terminated the call, put the phone in his pocket, and left the building.

This interaction points to a relatively tight integration between the man and his interlocutor. There was the informal and pre-configured opening as well as the immediate reference to a common task. With reference to humor, however, the conspiratorial laugh is perhaps the most interesting. The mission to Pentik's required only a quick introduction, as it drew on a common reservoir of knowledge. We can assume that both parties knew that the visit to Pentik's was on the agenda and so there was no need to unpack the information any further. The short laugh also pointed to a reservoir of solidarity within the dyad (Collins 2004a, p. 126). The laugh underscored that the negative result was perhaps to be expected, and this was simply and sardonically reconfirmed in the telephone conversation. The laugh was perhaps used to collectively shrug off the temporary problem and observe the shared estimation of the situation (Duncan 1970, p. 388). The interaction can be seen in terms of its ritual content. If the reading of the situation is true, the interaction was a minor ritual in which the paired identity of the two interlocutors was reconfirmed. The familiarity of the interaction described a well-integrated dyad. To the degree that it is possible to tell from the material, the two were mutually focused on the tasks involved in the maintenance of their coupled identity (Berger and Kellner 1964; Ling 2006). Others who had only partial access to their symbolic world would not have understood the context. They would not have had the familiar access to one or the other of the telephonic partners nor would they have understood the laughter in that particular situation. Long introductions and explanations would be needed to inform others of the nature of the errand and the reason that a laugh was in order when the store did not have the particular item for which they were looking. Finally, the event was not just a one-off situation. In a small way, it became a part of the couple's ongoing ballast. It probably was a reference point for future interactions. The reputation of the store as not having needed items might get played upon and embroidered in other situations, and the fact that the man got drenched in the rainstorm during his fruitless search might add some spice to the discussion the next time something is needed from that store.[33]

Humor and Power Relations   The use of laughter also helps to round off the coercive effects of group membership, which makes it an effective form of social control. Where group membership may require members to make sacrifices, a joking culture blunts some of the abrasiveness of these demands (Fine and DeSoucey 2005, p. 11). It helps, for example, to motivate others in the face of unfinished tasks. In the following observation we see the use of what Goffman called "bracket laughs"—that is, a laugh voiced by the person initiating the interaction used to mark the beginning of a lighter part of the interaction (Goffman 1981, p. 317).

Observation   A well-dressed man with a nice leather briefcase was sitting on a bus with a copy of the *Financial Times*. He discussed some business issue with a colleague. Eventually he began to conclude the conversation:

I thought the part you drafted fit in rather well (chuckle).
(response from interlocutor)

Yeah.
(response from interlocutor)

OK, then I will leave you to get on with it.
(response from interlocutor)

Yeah, bye.

In this case, the person being observed ostensibly had a common project with his interlocutor. It was not clear if the two were both employees in the same organization or were collaborators from different groups. The comments of the man underscored that some work remained to be done. The point of interest here is that he paid his colleague a complement that was punctuated with a type of Goffmanian bracket laugh. This led to an appeal to a sense of camaraderie that would eventually facilitate the completion of the work (Holmes 2006; Lerum 2004). This task done, the interlocutor sent the man off to work.

Given the direction of the compliment (from the co-present man to his interlocutor) and the ability of the co-present man to dispatch another person to continue with his work, we might suspect that the person on the bus and not his telephonic partner had the position of relative power in this situation. Still, a light laugh was used to gloss over any sense of coercion. "The creation of a joking culture," Fine and DeSoucey write, "serves as a strategy by which group members organize themselves and regulate their relationships by establishing a desire for cohesion." (2005, p. 7) This can be seen as small-scale ritual interaction.

According to Fine and DeSoucey (2005, p. 6), "the joking culture must be appropriate in light of the status system of the group. Ongoing joking differentiates group members, organizes status, and creates a social cartography of group life." Those in higher-status positions have a stronger hand when it comes to the development of a context where joking, banter, and laughter are appropriate. They have a greater prerogative to start repartee and to tell jokes. They can decide what is funny and what is not. They decide what is over the line of respectability, and they, more than others, decide what individual or group is a legitimate butt of humor (ibid., p. 16; Lampert and Ervin-Tripp 2002; Holmes and Marra 2002).[34] Often men have the prerogative to posit humorous statements, and often women are in the position of reacting to and supporting these statements.

Observation   Near the travel-information board in Oslo's central train station, a Swedish woman with a child in a baby carriage was engaged in a conversation with another person that appeared to be phatic. They were exchanging information and making small talk. The woman often interjected "Yeah!" or "So nice (laugh)" or "Super." In several cases, she engaged in interaction turns where she listened to her interlocutor, then made a quick comment followed by a hearty laugh. Examples:

(comment by interlocutor)
"I have to do that! (laugh)"

(comment by interlocutor)
"Yes, I really think so. (laugh)"

It seemed as though her interlocutor had posed the first part of a quip and that she was closing the interaction by making a quick confirmatory comment followed by a hearty laugh. There was never the opposite form where she provided the retort. While it is impossible to know, this style of interaction might indicate that she was talking with a man. Her comments were not overtly humorous; rather, she seemed to be confirming the comments of her interlocutor. This sequence was followed several times.

During the latter part of the conversation, her child had stood up in the carriage and the woman needed to hold the baby by the sweater with one hand to keep the baby from falling out of its carriage while using the other hand to hold onto her phone while talking. Thus, her attention was divided between keeping her child from falling out of the baby carriage and talking on the phone. There were several more quip/laughter interactions, but the phrasing of the interaction indicated that she was trying to close the conversation. Eventually she began the closing sequence and said "We have to get together. Good luck."

The telephonic partner seemed to have set the scene, and it was the remote partner who determined the pacing and the direction of the interaction. The woman seemed to be the more passive partner in the interaction and seemed to have the role of "straight man" (Sykes 1966). We also see a woman who is using laughter in order to manage the interaction. It has been noted in other places that women are often accomplished at using various communication devices (Fishman 1978; Treichler and Kramarae 1983). These include "back-channel" communication, such as interjections—i.e., grounding devices such as "mm" and "yea" and as is the case here, the supportive laugh—when engaging in a conversation (Sattel 1976; Clark and Brennan 1991; Clark and Marshall 1981; Clark and Schaeffer 1981; Duncan 1972; Johnstone et al. 1995; Kendon 1967). Linguistic strategies such as those noted here can underscore the hierarchical dimensions of an existing relationship (Tannen 1991). These strategies are used in the development and maintenance of communication channels (Imray and Middleton 1983; Gluckman 1963; Jones 1980; Rakow 1992). Despite the increasingly awkward situation with respect to her child, the woman did not break off the conversation. It was only after going through a longer closing routine that she was able to turn her attention to the child. The fact that the woman strove to maintain the telephonic interaction in spite of the emerging problem she had in managing her child underscores Goffman's sense that we have an implicit responsibility to the maintenance of the situation. In this case, there was a competition between the local and the mediated interaction where, in the short term, the mediated interaction won. In spite of the power issues that were being worked out here, there were cohesive elements (Lerum 2004). The nature of the repartee indicates that the woman was happy to spend her time in that interaction despite having to split her attention between the conversation and attending the baby. In ritual terms, repartee has perhaps the stronger claim on being seen as a ritual catharsis in the small-scale Goffmanian sense. The individuals who participate, and it is often just two, both share in the staging of the situation. Each of them draws on their wit and their reading of the situation to turn the discussion an interesting direction. If the two are able to bring off the interaction there is a tempo associated with the set-up and the come-back. There is a sense of mutual engagement and an entrainment in the interaction. The interaction eventually runs its course and the participants are left to find a new theme.

Humor and Repair Work   In addition to providing a catalyst for cohesion and a way to work out power issues, humor is a routine that is used to maneuver around difficult social shoals (Duncan 1970). When, for example, some sort of misunderstanding between friends has been uncovered, there is the need for a device that allows the individuals to put that issue behind them and to deal with the reframed situation. Laughter and humor can be drawn upon to disarm the situation. Laughter is a ritual process indicating that we are willing to let bygones be bygones. We see this in the latter part of the following observation, where some misunderstanding had arisen between a woman and her telephonic interlocutor.

Observation   A well-dressed woman came up an escalator in a large department store in London while talking on her mobile phone. She walked from the escalator through a clothing display area and across into an adjoining hallway. As she walked, she took part in this dialogue:

Exactly.
(long pause while listening to interlocutor)

But I thought he was *your* friend! (uttered in a somewhat sharp tone)
(long pause while listening to interlocutor; some grounding statements from woman being observed)

I saw him at a bar.
(long pause while listening to interlocutor; some grounding statements
from woman being observed)

He is so sweet.
(response from interlocutor)

Yeah, he is so sweet
(response from interlocutor)

Yeah. (laugh)
(response from interlocutor)

Yeah, he is so sweet. (laugh)
(The woman disappeared around the corner of the hallway.)

The laugh here seems to be a part of a repair project (Goffman 1981,
p. 118). The call—fitted into the time it took the woman to walk from
one area of a store to another—contained light gossip (the identification
of the man's allegiance); there was also the use of levity when the man's
status was established and a misunderstanding was clarified. There was
clearly some confusion as to whose friend the man was. After that was
sorted out, the woman reverted to more of a listener mode. In this part
of the interaction, the woman was using the routines of showing solidarity in her supportive responses. The laugh was a device that seemingly
marked the woman's acceptance of her interlocutor's version of the story.
Thus, it is tempting to think that the good humor was a way of marking a
successful interaction (Collins 2004b, p. 108).

In the preceding case, a laugh marked the conclusion of a process.
Laughter can also mark the beginning of a new process. In the following
example, the laugh seems to indicate that there is a new problem in the
interaction and that this new issue will necessarily mean more work for
the woman in question.

Observation   A woman was talking on a phone as she walked across the
waiting area of the Oslo train station. As I started to hear the conversation,
she was saying "I will come at about 22:30." She took some papers out of
her bag and read something to her interlocutor. She put the papers back in
her bag and continued to talk. It was clear that the other person had not
understood something so she took the papers back out and re-read them.
At that point, she gave a quick laugh and said "That means I have to go all
the way back again." This was followed by some more discussion. The

woman reversed her field and re-crossed the waiting area, returning to where she had come from. At the far end of the waiting area, she went up an escalator. When she was at the top of the escalator, she went to the cashier of an internet rental area in the station. She showed him papers and filled out a form. Eventually the cashier turned and spent a minute or two on his personal computer—it seems that he was recovering some information regarding the information that had caused the woman to return. He printed out a piece of paper and gave it to the woman. She took out her phone and made a call, presumably to the same person. She moved to a glass wall at the edge of the balcony and turned her back on the balcony area, shuffled through her papers, and knelt down so as to use the floor as something like a desk for her papers. She gave some of the information from the papers to her interlocutor and waited several minutes without saying anything. She then conversed with the interlocutor for a few more minutes, moving toward an elevator as she completed the call.

The woman may have been making some sort of reservation that she needed to coordinate with another person. The part of the conversation that was overheard indicates that there was some glitch in the order to which she needed to attend. The interesting thing in this context is that when the inconsistency was discovered, she added a laugh to the comment "Then I would have to go back again."

The laugh was a social lubricant indicating "This is a difficult issue, and it means more work for me." It is not a joke in the sense of "Have you heard the one about the old farmer?" Rather, it was a kind of salve to help the woman and her partner through an awkward strait in the conversation. As the misunderstanding concerning the papers arose, her laugh was as if to disarm the situation and say "We need to get this right." It is also interesting that this took place via mobile phone. The device allowed the partners to discover the problem and to clarify it in a quick cycle of microcoordination.

Another use of levity can be the repair of awkward situations. In the context of this book, it is tempting to see the awkward situations as examples of failed or at least strained rituals. Consider the following observation:

Observation   A middle-aged woman was walking quickly across an open area outside an office complex outside Oslo at approximately 8:55 on a weekday morning. She took her mobile phone and called a number as she walked and started the conversation by saying "Hi, this is Ragnhild. It was a little hectic this morning." As she said "Hi, this is Ragnhild," she had a slight chuckle in her voice. Perhaps she had had to drop off children at a day-care center en route to work, and the stress of getting to work had meant that this was done by cutting through some ceremony at the day-care center. She continued to walk quickly toward one of the entrances to the office building. As she walked, she continued to talk on the phone about afternoon plans. After about 30 seconds, she said "Goodbye" and hung up.

Although the entire situation was not available for observation, we are led to believe that the pre-work procedures of (for example) getting the children placed in their daily routines had gone quicker than normal and that the woman was on her way to a 9 o'clock meeting. This meant that, as she had a short period of time between the parking area and her office, she had a chance to do a small repair job. Two things are interesting here. First, mobile communication, along with a short period of "open" time, gave her the chance to mend fences. The communication channel facilitated the ability to call the "injured" partner in the interaction and to make amends. Instead of not doing the repair work, or waiting until she was in her office, she took advantage of the gap in her schedule to deal with the situation. The second issue here is her use of light humor in the greeting. Transcription of the opening phrase is difficult, since the rules of orthography do not allow the inclusion of words with a simultaneously voiced chuckle. The effect, however, was that in addition to introducing herself and giving a short summary of the recent incident, she also indicated openness to the situation. Rather than anger, there was an invitation to camaraderie.

That the call was made in a short span of open time not long after the incident had taken place and that humor was used in the opening show that the woman had understood the threat to propriety and was on the way to repairing it. In this process, she drew on a repertoire of effects that included timeliness and humor.

In the examples cited here, humor is a tool for the development and maintenance of cohesion within the group. It was used via mobile phone either to work on the creation of solidarity or as a repair strategy. The incidents cited here probably were not the only interactions between the interlocutors. Thus, the task of carrying on a friendship also took place in co-present situations. However, the use of a mobile phone extended the opportunities for the individuals to forge experiences and to carry out the minor—and in this case humorous—rituals of everyday life.

Marking the Boundaries (Sexual and Racial Humor) of an Out-Group Another side to humor is its use to mark social boundaries. This is seen, for example, in gender-specific scatological text messages that trade on references to genitals, sexual perversions, etc. There are also examples of text messages that play on racial stereotypes. Here humor is used to mark cohesion not by appealing to the common characteristics of the group members but by making reference to the differences between those who are members of the group and those who are not (Fine and DeSoucey 2005, p. 11). Humor can serve to mark the difference between the in-group and the out-group. Jokes that make other stigmatized groups the butt of the humor serves this goal. They can even be used as a kind of initiation ceremony wherein a person is not seen as a full member of a particular group until he or she is able to tell and laugh at this kind of humor (Goffman 1963b, p. 83). According to Fine and DeSoucey (2005, p. 9), "limits exist as to what group members can 'get away with' . . . often referring to the legitimacy of explicit references to race, intimacy, bodily fluids, or sexual orientation," and "members are expected to know these limits and to abide by them."

Humor can be used to underscore in-group/out-group differences by, for example, appealing to a person's sex, age, or cultural background (Duncan 1970, p. 256). So long as the people who participate in the joking or repartee share the same perspective, their comments can be at the expense of those "others," be they the "bosses," the immigrants, the men, the members of the other political persuasion, the snobby, the proletarian, the reborn, or the unrepentant (Hatch and Ehrlich 1993). The point is that humor, be it a joke, a quip, or repartee, often plays on the tension between groups and gives a group a simple way of crystallizing the absurdness of the "others" while providing a backhand celebration of their own correctness (Duncan 1970, p. 389). There is the danger, however, that humor of this kind can go over into derision and mockery. If that happens, the humor loses its ability to facilitate in-group solidarity (ibid., p. 404).[35]

■

Humor is ritual in which group solidarity is developed and maintained. When we tell jokes or engage in banter, we play into the mutual engagement in a mood that fulfills Durkheim's notion of effervescence. Joking and banter can touch on topics that are important in the self-definition of the group in that it can parody out-groups. In these small-scale Goffmanian situations, there is often a mood of levity, and there is engagement in the process. If the joke-telling session is successful, it contributes to the cohesion of the group by evoking issues that are central to the definition of the group. If the joke or the rejoinder falls flat, it can work against the integration of the group and can even lead to bad feelings. Joking and repartee can just as easily be carried out via mobile phone as in co-present situations.

Gossip as Ritual Interaction

The final form of ritual interaction to be considered here is gossip. Like the other forms of interaction, it often involves the definition of a closed group—in the case of the telephone conversation it obviously is a dyad. As we are drawn into the gossip session, there is entrainment and there is establishment of a common mood. As with other Goffmanian rituals, it is the gossipers who are responsible for the staging of the event. Mobile communication has provided us with a new stage upon which gossip can be carried out. Indeed, the point-to-point or more specifically the person-to-person nature of mobile telephony facilitates the mediation of gossip in the idle moments of daily life. This can be seen in the following observation:

Observation   A man dressed in a t-shirt, jeans, and jogging shoes was standing in the entry area of the Paddington-Heathrow train. As the train traveled toward Heathrow, the conversation proceeded as follows:

He is just that kind of guy.
(response from interlocutor)

Let me say it like this; I am not allowed to even call you.
(response from interlocutor)

Yeah, he's just like that.

Ambient noise kept me from hearing more of this sequence. Eventually I heard:

She is real hysterical, and Steve said she was like that.
(response from interlocutor)

Yeah.

The conversation continued in the same vein for 5–10 minutes.

Clearly, something important was at stake. The man I could hear was not holding back much in his estimation of the two people in question. There was a focused intensity in his comments. Since he had verbal room to address the faults of several people, it is easy to assume that his interlocutor was receptive to the interaction.

It is easy to tick off the dimensions of ritual interaction: a bounded group, a shared mood, and intensification of the interaction. In line with the idea of small-scale Goffmanian rituals, the two interlocutors were responsible for both the staging of and the participation in the interaction (Collins 2004a, p. 85). Finally, all of this was being done via mobile phone in stolen moments while the one person was on the train en route to the airport.

According to Jaworski and Coupland (2005, p. 691), "sharing gossip is a little ritual, a liminal state in which the participants bond in a state of communitas." In telephony, gossip has long been a part of the scene (Fischer 1992). Gossip is easily accomplished, and in some ways enhanced, when done via mobile phone. Indeed, some have suggested that gossip constitutes a considerable portion of mobile communication (Fox 1996).

Gossip and Social Cohesion   Gossip has been an object of considerable academic interest in psychology (Turner et al. 2003), anthropology (Gluckman 1963), and communications (Fox 1996). Goffman (1963b, p. 78) talks about the role of the stigmatized with respect to gossip in his book on stigma. He describes the small talk and gossip exchanged between the owner and the workers at a hotel in the Shetland Islands. He notes that the intimate interaction was dropped when guests were present (Goffman 1959, p. 116).

Gossip, in it its most innocent state, is simply talking about a third person who is not there (Turner et al. 2003). Going a bit further, gossip often includes speaking—or writing—value-laden information. And it can be used to impart critical or disparaging remarks (Fox 1996, p. 5).

The general consensus in the scholarly work is that gossip is seen as a way for a group to develop and maintain cohesion. The right to gossip is a sign of inclusion in the group (Gluckman 1963; Eder 1988; Davie et al. 2004).[36] When we are included in gossip, we have shown that we are trustworthy and that we have enough insight into the machinations of the group so as to have the ability to contribute to the interaction. Thus, the mapping of gossiping rights is the mapping of group membership.

Gossip is an interactive situation in which, if the interest of the partners is out of step, the entrainment is lost and the session is seen as a failure. Fox (1996) notes how important it is that both parties in a gossip session conducted via mobile texting adopt the same tone. The person receiving the tidbits must show the same degree of enthusiasm and animation as the person who is offering the information. If there is not a sense of reciprocity, the exchange is not a success (ibid.). The difference between men and women is particularly telling here. Fox reports the following sequence of comments:

Men don't get this, they don't understand that you're supposed to go 'NO! Really?!'

Yeah, with women it's always 'Oh My GOD!'

That's right. For women, gossip is a two-way thing.

Unlike many (not all) forms of humor, the willingness to share gossip transcends a kind of social boundary. Since it is potentially hurtful toward a non-present third person, gossiping is a slight transgression of normal social interaction that we should speak no evil of others. In gossip, and particularly in disparaging gossip, we cross over a threshold into a kind of mutual conspiracy to "share the dirt." The person with the latest news relates it while the other provides supportive "grounding statements" such as "I knew he [or she] was like that." There is in gossip, then, the experience of a slight mutual indiscretion. The gossipers share secrets. Thus, engagement in gossip is a way in which the cohesion between individuals is supported and developed.[37]

∎

As was noted above, gossip defines the boundaries of a group. According to Gluckman (1963), a strategy for putting a potential interloper in his or her place is for the routine members of the group to simply start gossiping. In the same way, an individual can pointedly refuse membership in a group by refusing to engage in gossip. Thus, when members of a group—particularly an in-group—start to gossip, there is pressure on others to join in (ibid.). To engage in the ritual of gossip—at least in its less salacious forms—is tantamount to a duty of membership (ibid.).

Gossip allows groups to work out shared value sets. Where everyday life presents us with complex and difficult situations (the reputation garnered by, for example, excessive drinking or sexual activity) through which we need to navigate, gossip helps to clarify these issues. It provides us with some insight as to how we might be viewed if we were to choose one or the other path (Jaworski and Coupland 2005, p. 690). The production of gossip (or collaborative narratives, as Eder calls it) means that the group collectively engages in the production of a story line and its implications. This account is a collective effort both in its construction and in its publication (Eder 1988, p. 226). Gossip simplifies and renders the dimensions of the group's ideology and morality and thus helps the group understand how to deal with intricate situations.

As probably was the case in the observation of the man on the train to Heathrow, in our gossip-laden remarks about others we assert our common values. A sense of ideology or principle is applied to a complex situation. In this process we also cultivate a connection with our interlocutor, perhaps at the expense of an absent third party. It is through these efforts that we feel engaged in the flux of the group. It is in this process that we gain a sense of solidarity (Jaworski and Coupland 2005).

When we gossip, we are also asserting a form of social control. Eder (1988, p. 226) discusses how teen females use gossip to regulate status ambitions within the group. In the use of gossip, those who are seeking expanded popularity become the focus of gossip. This can serve either to draw them back into the sphere of the group or to definitively exclude them. Thus, there is also the exercise of power in gossip. The people trying to assert their priority over the others can be sabotaged by the gossip and the scandalmongering of the other group members.

The power dimension is also seen in the way that retribution is managed. There is always the danger that if today's gossipers transcend some boundary, however this is defined, that they will be the victims of tomorrow's tittle-tattle.[38] To avoid this, the gossipers need to feel secure in their partners and they need to take care that they do not extend the gossip beyond a certain level of detail. While there is often conjecture in gossip, we need to rein in the worst speculation. A gossiper who carries speculation too far becomes labeled as a gossipmonger (Gluckman 1963). A little gossip engenders a sense of common insight and solidarity, but too much allows others in the group to paint the picture of a tattler who is not to be trusted. These characterizations are worked out in the context of the in-group power structure. Thus, there is a balance in the use of gossip. It has useful functions in terms of in-group cohesion, but it is traitorous to tell the in-group narrations to persons who are outside the group. Internal gossip is the sign of a well-functioning small group. Gossip tells us that there is trust between group members, that the power relations have reached some equilibrium, and that there is the ability to define a common ideology or ethic. At the same time, if the loose talk is repeated to those who are not seen as being inside the circle this is a betrayal of trust. According to Gluckman (1963, pp. 313–314), "when you gossip about your friends to strangers . . . you are admitting them to a privilege and to membership of a group without consulting the other people involved."

Gossip and the Mobile Phone   Perhaps more than with humor, the mobile telephone influences the process of gossiping. As is obvious from the observation of the fellow on the Heathrow train, the mobile phone is a fertile technology with which to foster gossip. Gossip by telephone— both land-line and mobile—has most commonly been treated in gendered terms (Marvin 1988; Ling 1998; Kearney 2005; Fischer 1992; Aronson 1977; Rakow 1992). Marvin (1988) talks about women's use of the phone for gossip as a common stereotype.

Mediated gossip allows for the development of in-group cohesion and is a channel where the individuals can engender the appropriate mood by using voice inflection in voice interaction or by using all caps or multiple exclamation points in texting or instant messaging (Ling and Baron 2007). In the process of these sessions via the mobile phone we develop and elaborate in-group values and we work out the power relations within the group. This characteristics of gossip means that the mobile phone is a perfect device for tucking gossip sessions into the folds of daily life (Fortunati 2000). It is a channel through we can directly reach a specific person, and thus it creates an omnipresent direct connection between individuals. In this way, it is amenable to the exchange of juicy tidbits about other persons whenever and wherever the gossipers might find themselves.

∎

The sense that has perhaps been developed here is that gossip in its most spicy form is a context in which we best see ritual interaction. It is also clear that the mobile phone can be used for gossip of this more caustic

sort. The same results are reported by Fox (1996), one of whose informants said: "I've got a very close group of girlfriends and we know all about
each other's lives. Yesterday on my way to work my friend called me on
her way home from a one-night stand." It is clear that in both cases there
were a lot of thoughts and feelings in motion. It is also true that when
the talk—or the text—takes a particularly strong turn there is the greatest
buzz. The man on the train who was trashing various other people and the
woman who reported her recent nocturnal activities were engaged in this
kind of effervescent talk .The man on the train was so engaged that it was
easy for other passengers to follow along from several yards away. In that
case, the mood and the intensity were in place and the evidence suggests
that there was an agreement as to the values being expressed.

Gossip, however, does not have to be vicious or intensely felt. Simple
chit-chat in which the interlocutors update each other on recent events
and the situation of other common friends can serve as a kind of phatic
interaction. In this case the point is not to exchange groundbreaking
information or even to carry out any instrumental interaction, such as
agreeing on a time to meet. The point is simply to "touch bases." In this
case, gossip is simply a phatic device by which telephonic partners simply
assure one another that they are still active friends.[39] Here is an example
of such interaction:

Observation   A woman was walking along Regent street in London while
talking on a hands-free phone. The conversation proceeded as follows:

Sorry, I am on the street. It is noisy, it is so noisy.
(response from her interlocutor)
(unintelligible talk from the woman)

When is Sally done with work?
(response from her interlocutor)

The woman approaches a street crossing and describes her situation further:

Run, run, run, I never wait for the light when crossing (laugh).

This was followed by an extended discussion about jaywalking.

How was your vacation?
(extended discussion about vacations)

The woman disappears into a store.

There were no pointed barbs or revealing comments. The woman was
engaged in simple chit-chat. The talk shifted from one theme to another:

when Sally would be done working, the prevalence of jaywalking, summer vacations. There was a lot of laughter, and themes shifted quickly. The talk was supportive and perhaps aimless (Tannen 1991; Rakow 1988). Thus, the talk did not necessarily invite collusion in the sense that the talk of the train gossiper or the woman reporting her new romantic adventure. It was, however, making contact and filling one another in on the details of everyday life.

We can ask if the aimless chit-chat of the woman on Regent Street constituted gossip. Although a third party was discussed in the conversation, there were ostensibly no value judgments being made. If this was gossip, it was mild gossip. Nonetheless, it illustrates a common use of the mobile phone. Fox reports that about half of all calls included various forms of chatting. Though not all of this was gossip, there is a sort of talk that, in a mild way, helps to form the trust and the context for other types of exchanges (Fox 1996). As we see from the observation of the woman walking along the street, the device provides a direct channel to other members of the group that can be used during the open moments of the day. We can get a quick update while waiting for a bus or an appointment or while simply moving from one engagement to another. If it is inconvenient to talk, we can use the texting function to exchange information.

The material here shows that gossip is a form of ritual interaction that is staged by the participating individuals, draws collective attention of the people who are participating, demands a common mood, and results in a mutually understood sense of participation. Most importantly for the analysis here, gossip can be carried out by mobile phone, and indeed the use of a mobile phone may actually enhance the ability to gossip. When one is using a mobile phone, barriers to non-members are easily erected via either the choice of location or the use of texting.

Conclusion

There are various situations in which the use of a mobile phone seemed to tear at the fiber of society. While the individual is perhaps well ensconced in a telephonic interaction, the effect on the local situation can mean that the co-present individuals have to make room for the actions of the telephonist. That is one side of the equation.

However, mediated interaction also plays into the ritual form, and thus it actually enhances the cohesion of the group, be it a dyad, a small group of teens (Ling 2005a), a church prayer group (Rafael 2003; Palen et al. 2001), or attendees at an Alcoholics Anonymous meeting (Campbell and Kelley 2006). I have examined situations such as greetings, or perhaps our willingness to dispense with them, the negotiation of romantic interaction, texting and the use of various forms of argot, humor, and repartee, and the use of gossip. I have outlined how the mobile phone affects how groups are constructed and how they are brought together in the interaction. The mobile phone is often adopted by pre-existing groups that were formed in other contexts. Thus, there is often a pre-existing solidarity or internal cohesion within the group. What does the introduction of the mobile phone mean in this context? The assertion here is that the way ritual is performed is carried over into the mediated interaction of the group. In some cases the interaction is modified and even strengthened. The lovers, the people telling jokes, the people exchanging gossip, and the people who use pre-configured greetings in general have a background that is established in co-present interaction. In these cases, however, the mobile phone has extended the range of the group interaction. The lovers can send "sweet nothings" to each other through the day and, in a small way, rekindle the mood of the last date. The friends or work colleagues can use humor in their interactions in order to get around difficult situations, to motivate the other person, or simply as a phatic device to let them know that they are valued. The friends can exchange gossip, either over the back fence or via the mobile phone, and in the process reaffirm moral standards of the group. Finally, pre-configured greeting sequences (or, in one of the cases cited here, picking up the thread of an existing interaction via the use of the mobile phone and the caller ID function) mean that the interlocutors maintain the same line of interaction as they carry out their alternatively separate and collective Saturday morning errands. The use of mobile communication influences the ongoing ritual interactions by giving these people another channel through which to interact.

In texting and in the use of jargon by teens, ritual interaction seems to be more uniquely related to mobile communication and the possibilities allowed by texting. The use of the z endings exists almost exclusively in the world of mobile phone texting. The use of faux Swedish also seems to exist there but not elsewhere. This points to the possibility of uniquely mediated forms of ritual. These formulations are obviously commented on when the teens are co-present. For example, they discuss how "out" it is to use the z formulations and how funny or positive it is to write the quasi-Swedish words. In this respect, the co-present interaction supports the mediated culture, and not the opposite (as in the case of romantic interaction).

Mobile mediated interaction has the potential to create group cohesion. As when Durkheim sees the group experiencing a mutual sense of themselves when they are "shouting the same cry, saying the same words, and performing the same action" (1995, pp. 231–232), the list can be extended to include joking, gossip, romance, and argot via the mobile phone. Arminen (2007) points to how mobile mediated ritual supports social cohesion:

Chains of connotative meanings get established in this manner, and the coordination of social action intertwines with the symbolic organization of everyday life, establishing the actor's habitus that signifies the chosen way of life. In the ubiquitous mobile presence, the settings and activities may get symbolic embellishment. Seamless communication not only rationalizes time usage but intensifies social presence. The accountability of action extends both to the timing and social dimension of activities. Instead of accountability afterwards, the perpetual contact extends to the very moment of accomplishment of activity. Intimate connection to everyday life also enables social accountability of actions and choices through seamless communication. Depending on the social configuration, this extended accountability may strengthen external purely goal-rational control as well as social responsibility.

Like Licoppe (2004), Arminen is saying that the mobile phone extends the opportunities that we have to know each other. Connected presence is a construction that is contrasted with more traditional interaction between friends where there are relatively long periods of no contact punctuated by short "get-togethers." During the time where they are co-present the friends catch up on the main events in their respective lives. Licoppe contrasts this with connected presence that is carried out via the mobile phone. Since the threshold for interaction is lower, the members of a friendship circle can freely contact each other whenever and wherever the mood moves them. They can call or send text messages when the impulse strikes. This means that the "get-together" takes on a different tone. Since all are relatively updated due to the background traffic, the co-present sessions do not have the function of allowing the individuals to catch up. Rather, the "get-together" allows them to carry certain lines of narration further. Thus, the device is useful in extending the original bond beyond co-present situations into the folds of daily life. In slightly different terms, the interaction, both mediated and co-present, can be more effectively used to develop the ideology of the group.

The mobile phone extends the ritual reach of society beyond co-presence. Thus, the device extends the reach of parents, children, and friends. Rather than relying on a totemic representation of a social circle, they are, to the use phrase of Katz and Aakhus, in perpetual contact (Katz and Aakhus 2002a). The point here is that the mobile phone seems to result in stronger internal group bonds. This idea seems to fly in the face of the work of Putnam (1995), who fears for the reduction in social capital due to mediation. It also seems to contradict the suggestion by McPherson et al. (2006) that we have fewer and fewer confidants.

9

Bounded Solidarity: Mobile Communication and Cohesion in the Familiar Sphere

A recurring theme of this book is that ritual interaction is one way in which we develop social cohesion. The previous chapters have outlined the development in thought regarding ritual interaction and how it influences the use of the mobile telephone.

It is in continual everyday ritual interactions that our sense of being social is developed and sustained. We continually draw on these when we interact with other people, both those we know and those who are strangers. We have a repertoire of devices which we draw upon as we deal with the ebb and flow of daily life. We can greet a friend with familiar phrases and gestures, just as we can go through the process of buying a newspaper when a stranger is behind the counter of the newsstand. We can talk with a close friend about the state of his or her marriage, and we can make small talk with a stranger about the weather. The point is that we—as well as our counterparts—have the social background that allows us to master these situations. Seen from the perspective of ritual, they are events where the participants share a mutual focus and a common mood. The events play on pre-existing forms of interaction; they also help to produce—at one level or another—a form of cohesion.

The mobile telephone is the tool of the intimate sphere—that is, the sphere that includes our familial and romantic interests. But, perhaps even more important, it the sphere of interaction that includes the closest friends. In chapter 7 we saw that it does this perhaps at the expense of the non-intimates. While generally we must be open to both intimates and strangers when we interact in daily life, the mobile phone tips the balance in the favor of the intimate sphere of friends and family. In a situation where there otherwise might have been the opportunity for talking with a stranger (e.g., waiting for a bus or standing in a checkout line), we can instead gossip, flirt, or joke with friends, intimates, or family members. It allows us to pick up the threads of ritual interaction when and wherever the urge moves us. This is, of course, the other major point. The device makes us individually addressable. Instead of calling to a geographic location in the hope that our intended interlocutor is nearby (as with the land-line system), we call or send messages directly to the individual.

The ability to sustain social contact is basic. As I noted in chapter 7, we can even sneak interaction with friends into the flow of other local affairs. When I recently gave a talk to teens at a high school in Oslo, about halfway through the class period I asked how many of them had received text messages during the class time up to that point. Many of the students raised their hands. When asked what the messages were about, one girl said she was busy arranging a meeting a friend in the library after class in order to work on a project. A boy and a girl in the class had sent each other messages having to do with that day's assignments in the class, which papers were due, and the status of a common project. Perhaps most interestingly, a girl said she had received a text message that dealt with the planning of a visit to a movie theater that evening with her friends. I asked if they were working out the specific time that they would meet. She answered that they were working out not only the time but also which movie theater they would meet at. Thus, they had decided on a film, but since it ran at different places at different times they were working out the logistics of both time and place.[1] All this was being done discreetly in the middle of my presentation. The social life of the teens was being carried out in a discreet back channel. I had not heard a single peep or blirp from their mobile phones, but the activity was there none-theless, being neatly carried out in what Fortunati calls the folds of life. Some of the traffic dealt with educational issues, some not. Some could have been dealt with via old-fashioned notes, but most of the messages assumed that the interlocutor was beyond the reach of these. The teens were shuffling between two interactions, namely the negotiations with friends and attending to my lecture. Both interactions were taking place simultaneously, and each of these interactions had its form and meaning in terms of the future functioning of the social group.

The point is that mobile communication is a part of the situation for these teens. They were engaged in the maintenance of their social inter-actions. To use the phrase of Katz and Aakhus (2002a), the teens were engaged in a kind of perpetual contact that particularly favored their friends. It was not being used as an open-access system, such as PC-based instant messaging or social networking. More than anything else, it was being used as a medium for the familiar sphere.

. This was a very Goffmanian moment, albeit on a dual stage. On the one hand, the students were doing an excellent job of appearing to be students. They all had their pens and notebooks out. They were noting down various salient points from my talk and they all had the facade of being interested and engaged—at least that is what I would like to believe. My question about their use of SMS, however, showed that they were at the same time managing the interactions in their various social circles. Messages were being sent and received, meetings were being agreed to, and social events were being negotiated all the while. Translated into the discussion here, social cohesion was being managed in real time, both co-presently and via mediated interaction. The students were able to balance the here and now with the there and now. Ito and Okabe (2005, p. 271) comment on the same conflation of local and remote when discussing the use of mobile communication in Japan: "Keitai email [mobile texting] constructs a space of connectivity that relies on a pulsating movement between background and foreground awareness and interaction as people shift from lightweight messaging to chat to flesh meets." Unlike the Durkheimian model, where there are others who organized specific ritual events and who also take on the role as keepers of the paraphernalia, these teens were doing it all themselves. They were working out the social interactions with the proper tone and use of slang while also appearing to be diligent students sopping up my bits of wisdom.

I bring this up because one of the general findings from several analyses is that the mobile telephone tightens the interaction within the small intimate sphere. As these students were working out their social affairs, they were also in the process of building up what Berger and Kellner (1964, p. 9) call the nomos of the group. There is obviously the need for the social group to build up and objectify the meanings and ideas of the group such that they become a kind of common property. For example, the girls who were negotiating the movie were perhaps having to weigh decisions as to whether the showing at 19:00 at the Saga Theater would allow them time to meet beforehand at a favorite café and would allow Tina time to participate in band practice. If they took that alternative, however, it would mean that Kristin would have to rush home afterward, since she had a curfew. Thus, for Kristin's purposes it would be better to see the film at 18:30 at the El Dorado, but that was too far away for Kari. Weighing such alternatives and working out whose interests are taken into account are the ways in which the common meanings of the group become "massively objective."

As I will examine in chapter 10, such considerations become the common ideology of the group (it is better, for example, to take into account Tina's interests this time since she didn't get to go last week, etc.). In a broader way, these decisions, regardless of the medium in which they get worked out, are part of a group ethic or an ideology. It might be that the group has the general sense that staying updated on the films being shown is an important element just as it is more important to include Kristin than Kari in all events since she is seen as being funnier and easier to get along with. That is, there are fundamental elements in the ideologies and there are situations where they are worked out in practice. These ideologies are developed and nurtured through various forms of ritual interaction. They imply a certain discipline and they become included as a part of the lore of the group. The teens here might remember the time they wanted to see the new Johnny Depp film but Kari became frustrated with ordering the tickets and made the others uncomfortable with her comments about another friend who was not there. Through such interactions, the sense of the group is worked through the development of a common version of the world. Berger and Kellner (1964, p. 11) note that "every social relationship requires *objectification*, that is, requires *a process by which subjectively experienced meanings become objective to the individual and*, in interaction with others, *become common property and thereby massively objective.*" The mobile phone is a medium through which these ideological elements are developed. To take it one step further, the mobile phone means that this kind of interaction takes place on a nearly continual basis. Indeed, as I found out, the classroom is not spared from these machinations. Since we are always reachable, these interactions need not wait for the next "flesh meet."

In the previous chapters, I traced in some detail the use of different forms of interaction (the joke, the greeting, the gossip, etc.), each of which was a device that allowed the telephonic group to further draw in the circle. I was interested in examining the ritual tools and devices used to encourage social cohesion within the group. The broad conclusion is that the mobile phone produces some turbulence with regard co-present activities (chapter 7) but there are also ritual mechanisms as well as nearly perpetual accessibility through which the group is able to develop and elaborate social cohesion (chapter 8). In this chapter, I turn to the results of larger-scale studies that have also examined this issue, albeit from a more abstract perspective.

Mobile Communication and Integration of the Small Group

The assertion that the mobile phone is an integrating factor in the small group seemingly runs counter to the broader sweep of social interaction. In chapter 2 I described the reduction of social cohesion in society and its replacement with various forms of individualism. Putnam (2000) traces the decline of social capital in parent-teacher associations, bowling leagues, and other organizations. McPherson et al. (2006) documents the decreasing number of persons in whom we can confide personal matters.[2] Nie et al. (2002), discussing the internet, describe the re-prioritizing in favor of distant and more diffuse ties over those that are near and perhaps more meaningful. In each case, the studies indicate that the balance between being socially and individually oriented is tipping in the direction of the latter. Although these studies were based on the scene in the United States, and although there are countervailing trends, their general finding is that the social fabric is not what once it was. It is perhaps easy to simply say that it is an artifact of the living in the United States, but the findings also point to the drift toward individualism that has also been noted in other places (Beck et al. 1994; Beck and Beck-Gernsheim 2002).

The mobile telephone is not the solution to the diffusion of social interaction. Nonetheless, it does seem to have the potential to connect individuals in ways that are not possible for other technologies. Thus, while I concur with Collins (2004a) that social cohesion is best spawned in co-present action, I assert that mobile communication is a medium through which we can develop and maintain social groups.[3]

■

Analyses of mobile-based social interaction contradict the idea of social decay, at least in the context of the small group. In many cases, the research is showing that in small groups use of mobile communication technologies is actually fostering internal cohesion. In their summary of mobile communication Castells et al. (2004) say that, although groups are often born in co-present situations, they are reinforced via wireless interaction.[4] This assertion is reminiscent of the gossiping, joking, and romancing that were described in the previous chapter. In co-present interaction, the group produces collective practices that are, again, raw material for social cohesion, which is at least partially maintained in mediated interaction (Campbell and Russo 2003, p. 329; Haddon 2004, pp. 81–82).

A study of teens in Norway[5] found, among other things, that the intensity of both voice telephony and text messaging covaries with the teens' sense of social inclusion among peers (Ling 2005a). That is, the more a teen in Norway used his (or her) mobile phone, the more time he spent with friends, the less time he spent at home, and the less likely he was to report feeling lonely.[6] A more recent study in Norway indicates that those teens who use the mobile phone for voice calls also more frequently report that they feel themselves "more popular among peers"[7] and that they "have many friends."[8] The teens who reported many voice calls via mobile phone were not in agreement with the statement "I think it is difficult to make friends that I can really rely on"[9] (Koivusilta et al. 2005; Punamaki et al. 2006).

Material from Korea paints a similar picture. According to Kim et al. (2006), mobile communication channels reinforce strong social ties.[10] Kim et al. used a national sample of individuals (1,039 cases) and social network analysis. Their work indicates that the mobile phone is largely used for the mundane interactions associated with intimacy and friendship. They note that the mobile phone "seems to be used as the best channel for maintaining such strong relations" (ibid.). This finding is supported by material from Taiwan. According to Wei and Lo (2006, p. 68), the mobile phone is a device that allows the individual to strengthen the bonds to their family and expand their "psychological neighborhoods."[11] Wei and Lo note that the device "facilitates maintenance of proximity," particularly in the case of women. Finally, examination of material from the broader European scene again points in the same direction. An analysis of data from a broad survey of persons in Europe (Ling et al. 2003)[12] indicates that use of the mobile phone covaries with various types of informal social interaction. The data shows that the mobile phone was used to organize activities such as informal café visits and other social interaction with the small cadre of friends and family.

The persons who are contacted are often limited to the nearest family and friends; that is, they are close in both the social and the geographic sense of the word. Matsuda (2005, p. 127) examined the use of mobile communication in Japan and found that the device is used most often for contacting the individual's partner and after that, it is used to contact friends. In addition, Matsuda found that the ability to maintain contact via mobile phone—and before that the pager—served to stimulate physical meetings. (See also Dobashi 2005.) Smoreda and Thomas's work supports this general direction based on an analysis in nine-country European survey carried out by Eurescom in 2000.[13] They found that, although the traditional land-line phone was a strong element in family and friendship circles, use of a mobile phone was particularly strongly correlated to participation in friendship circles (Smoreda and Thomas 2001, p. 5).

Material from Sub-Saharan Africa points to some of the same findings, albeit with some slight twists. Donner (2005) report that there, as in other parts in the world, the mobile phone is often used to enhance communication within the group or the family. However, the device is also used as a "job telephone" to a much greater degree. Donner states that "microentrepreneurs[14] use their mobile phones to increase the frequency of their contact with friends, family, and existing business contacts, and to facilitate new contacts with business partners, suppliers, and customers" (2005, p. 28). While there is an entrepreneurial motivation to have a mobile phone, he reports that the major effect of the device is to increase communication within the intimate sphere. It also has the ancillary effect of expanding the individual's contact net and bringing him or her into contact with potential customers and suppliers. Slater (1963) also describes the use of mobile communication Ghana where a significant use is "funeral traffic," an important part of the social life in that country.[15]

The circle of persons called via the mobile phone is often relatively small. Based on her research in France, de Gournay (2002) suggests that the mobile phone encourages the scaling down of diffuse social relationships into a group of relatively close friends. Material gathered in Norway in 2002 shows that, although people have about 50 names in the name register on their phones, they contact only about 2 people on a daily and about 6 on a weekly basis. The material also shows that it is young adults (aged 20–24) who are the most socially outgoing, having a mean of 104 names on their phones. The material shows that on average they contact 4 of them daily and 10 of them weekly.[16] Ito and Okabe (2005) report generally the same findings for Japan. They write: "In our interviews with heavy users of [mobile phones], all users reported that they were only in regular contact with approximately 2–5, at most 10, close friends, despite having large numbers of entries in their [mobile phone] address books." (ibid., p. 264) Matsuda (2005, p. 133) calls this the "full time intimate sphere." Habuchi (2005, p. 167) writes: "The keitai can serve as a means of maintaining existing relationships when it is used to strengthen ongoing collective social bonds. Keitai do not allow entry of strangers into such collective cocoons. . . . There is a zone of intimacy in which people can continually maintain their relationships with others who they already encountered without being restricted by geography and time; I call this a telecocoon."

There is a tendency to use a mobile phone for communication within the closer social spheres. Family and even more so friends are the common interlocutors. Further, the people with whom we are in touch are often near at hand. Following from data gathered in Norway in 2006, about half of the calls made are to persons who are within 10 kilometers of the caller.[17] Material from the United States[18] shows some of the same though a bit more diffused. About 70 percent of calls are made to person within a 25-mile radius.[19] Interestingly, this analysis shows that those people who felt that the mobile phone was most important for coordinating family activities—about 40 percent of the sample—also had the largest proportion of calls to persons within 5 miles.[20]

■

There is perhaps an even stronger sense that texting strengthens the bonds in the immediate circle of family and friends. According to Castells et al. (2004), texting acts as a catalyst for the construction and reinforcement of peer groups. Similarly, Miyata (2006) notes that mobile texting supports interaction among friends and kin.[21] This notion is taken further in the work of Reid and Reid (2004), who find that a preference for texting corresponds to a preference for smaller tighter social groups.[22] Using material from an internet-based study that included largely persons from the United Kingdom but also a number of persons from other countries, Reid and Reid note that the social interaction among those with a preference for texting is "well-defined and close knit."[23] There is the assertion that texting is used to engage in extended exchanges that include turn taking and approach the dynamics of a voice-based conversation. According to Reid and Reid, "texters were more likely to text a particular group as opposed to many groups, and more frequently participated in several simultaneous text conversations, findings which taken together reinforce the idea that texters share interconnections within a close group of friends in perpetual text contact with one another" (ibid., p. 5).

Using longitudinal data, Igarashi et al. (2005) found that young Japanese build strong ties through the use of texting.[24] They also found that those who communicated both face to face and by mobile texting were more intimate over time than those who used only face-to-face interaction. Thus, there is a complementary effect between co-present interaction and texting. The one plays on the other to increase the degree of intimacy experienced by the individuals. Further, Igarashi et al. found that persons who use mobile texting were more likely to form cliques (groups of three or more individuals who were all in contact with all the other members of the group).

In the United States, Campbell and Kwak (2007), using the data set mentioned above,[25] examined the interaction between texting and calling on the one hand and informal co-present interaction on the other. They found that both texting and voice interaction predict informal socializing for those who live within a 25-mile radius. Thus, although there is a role for voice interaction given the special rate system[26] in the United States, texting is also a significant factor when it comes to connecting local peer groups—or at least those who are within driving distance (ibid.).

Texting has also developed into a channel where some groups feel comfortable expressing thoughts and feelings that were once reserved for co-present interaction. Data from Norway shows that 29 percent of teens and almost half of group those who are between the ages of 19 and 24 prefer texting over face-to-face communication or voice telephony when they have to communicate something that is uncomfortable. This compares to only about 3–4 percent of those in the 45–54 age group. The material also shows that 34 percent of all persons in the 19–24-year-old age group think that it is easier to express their feelings via SMS than through either voice telephony or face-to-face. By comparison, only 3.5 percent of the 55–64-year-olds think the same.[27]

In Japan, according to Ishii (2006, p. 360), "mobile mail [texting] appears to support only a closed network, whereas PC e-mail was found to promote friendship with distant friends."[28] Ishii goes on to say that texting supplements face-to-face interaction. It is not a channel through which new connections are made, but rather is supports pre-existing networks. Similarly, Smoreda and Thomas (2001, p. 5) noted that in Europe the use of SMS was more related to friends and in particular friends who were based relatively nearby. Ito and Okabe (2005, p. 263) note that texting is a form of communication that can be used almost regardless of location and the social constraints associated with location:

> What is unique about mobile text chat is the way it is keyed to presence in different physical spaces. We observed mobile text chat in diverse settings: home, classrooms, and pubic transportation. Like internet chat and voice calls, mobile text chat can be used whenever two parties decide to engage in focused "conversation." What is unique to mobile text chat, however, is that it is particularly amenable to filling even small communication voids, gaps in the day where one is not making interpersonal contact with others. . . .

Ito and Okabe go on to describe how the somewhat asynchronous nature of mobile text allows for pauses in the interaction in a way that is awkward in voice interaction. There is the sense in textual interaction that the interlocutor may have other affairs to deal with—for example, getting on and off trams, or dealing with the more pressing inquiries of his or her teacher.

At the same time, the channel is open. Indeed, it may never be closed. As Reid and Reid note (2004), the users of texting report that it allowed them to develop deeper relationships with their interlocutors. Individuals form what Reid and Reid call text circles, or tightly woven groups of texting mates who are in frequent contact via SMS (ibid., p. 7). Reid and Reid assert that the content of the text messages is not instrumental interaction but is rather more expressive and phatic in nature (ibid., p. 7; see also Johnsen 2000). The point, according to Ito and Okabe (2005, p. 264), is not to transmit information but to check on the status of the relationship:

> These messages define a social setting that is substantially different from direct interpersonal interaction characteristic of a voice call, text chat, or face-to-face one-on-one interaction. These messages are predicated on the sense of *ambient accessibility*, a shared virtual space that is generally available between a few friends or with a loved one. They do not require a deliberate opening of a channel of communication but are based on the expectation that one is in "earshot." . . . As a technosocial system . . . people experience a sense of persistent social space

constituted through the periodic exchange of text messages. These messages also define a space of peripheral background awareness that is midway between direct interaction and noninteraction.

In this mode, texting is Ito's "tap on the shoulder." A text message can be sent or received when and wherever, regardless of our co-present situation. It can remind us of our other social interactions. The social relationships in all their massive nature are continually available.

The material reviewed here suggests that we are in touch with a relatively small and perhaps selective portion of our social network on a regular basis.[29] Putting this together with the material from the previous chapter, we can see that the small ritual devices of humor, gossip, flirting, argot, and pre-configured greetings encourage a sense of familiarity that is easily extended into the world of mobile communication. The results reviewed here seem to confirm that these ritual forms of interaction are having their effects. Making a mobile phone call is a direct communication with the individual. While the device is not as suited to assisting in the maintenance of weak-tie relationships, it may be better than other forms of mediated interaction for maintaining strong-tie relations (Miyata 2006). Calling or texting a person's mobile phone is a very concrete application for their attention. By doing so we are inviting ourselves into their personal sphere.

The mobile phone makes us individually addressable. We are not calling a location in the hope that the individual will be nearby; we are calling a particular person regardless of his or her current situation. To make a call to a mobile phone is to personally address the individual. More than instant messaging, a visit to Facebook, e-mail, or land-line telephony, the mobile phone is an instrument of the coterie.[30] It has been shown to facilitate communication in the familiar sphere just as it has lowered the threshold for contact among the in-group. These findings may point to the suggestion that mobile communication encourages a kind of intense bounded social capital perhaps without the corresponding weak links (Ling 2007a). While the mobile telephone may be fraying the fabric of some co-located social interactions, it seems to be supplementing it in others.

Connected Presence

The material presented in chapter 8 and in this chapter indicates that mobile communication encourages greater cohesion in the familiar sphere by allowing for different types of ritual interaction. Looking further into how ties are established and strengthened in small groups, let us turn to how mobile communication has established itself as a medium for interaction.

The sociologist Christian Licoppe describes a transition associated with the adoption and use of mobile communication. He notes that the threshold for contacting one another has been lowered dramatically:

First, the mobile phone is portable, to the extent of seeming to be an extension of its owner, a personal object constantly there, at hand. Second, as there is no mobile directory, one gives one's number to a chosen interlocutor (Licoppe and Heurtin 2001). Explicitly inscribed in an economy of gift and counter gift, the ritual exchange of mobile phone numbers represents, for the protagonists, their entry into a mode of access of which the frequency and continuity of calls are no longer limited by access to localized communication tools. Third, the mobile phone includes an extensively used, portable, condensed list of the most important and frequently called numbers. (Licoppe 2004, p. 146)

Licoppe suggests that, at least in part because of mobile communication, we are shifting from a kind of telephonic interaction that was punctuated by longer interludes into what he sees as a form of more or less constant interaction. In the pre-mobile mode of interaction (that is when interaction was based on geographically fixed telephonic interaction), conversations were infrequent and often were more extensive in terms of the themes discussed. When the interlocutors called, they set aside time in order to converse and to review a series of themes. The image is that of two old friends catching up on each other's lives, or the obligatory Saturday call between an elderly parent and a grown son or daughter. It is a way of vicariously maintaining a relationship interspersed between face-to-face meetings (ibid., p. 139). In this mode, the interaction "consists of open and often long conversations in which people take the time to talk. These calls are made at appropriate times. Taking the time to speak on the telephone is read by participants as a sign of their mutual engagement in the conversation." (ibid., p. 141). As an illustration of this kind of interaction, Licoppe reports: "The telephone then becomes a resource for maintaining and nurturing the link, as in the following example of a student: 'If I don't call them for a week [my parents] worry about me and phone me, so I never forget . . . or rarely.' " (ibid., p. 143) The interlocutors budget the time to talk and—beyond perhaps the use of an opening gambit used as a reason to get in touch—the interaction wanders from theme to theme. The interlocutors might review the situation at their job or their progress on some larger project. They might review the achievements of children or spouses or they might exchange the latest news with regards the local goings on. There is the updating of each other on recent events both large and small and the emergence of new themes as the discussion progresses. These interactions have an open format, and there is room in the conversation to allow reciprocal turn taking.

■

The alternative form is what Licoppe calls "connected presence" (Licoppe 2004; Rivère and Licoppe 2005). In this form of interaction, as opposed to the languid interaction described above, the interlocutors engage in a series of short, frequent calls that may have an instrumental function but may also simply be phatic (Campbell and Russo 2003). Connected presence, according to Licoppe (2004, p. 141), "consists of short, frequent calls, the content of which is sometimes secondary to the fact of calling. The continuous nature of this flow of irregular interaction helps to maintain the feeling of a permanent connection, an impression that the link can be activated at any time and that one can thus experience the other's engagement in the relationship at any time."

In the mode of connected presence, the calls and messages are so engrained in the normal progression of daily interaction that they become normalized. If, for example, the group of teens is to meet at the usual place en route to school, the calls and text messages leading up to their physical meeting might be to confirm the various estimated arrival times. Another series of calls or messages might be associated with reminders regarding homework, coordination of clothing, or progress on a common project, and yet another thread of interaction might be to express mortification or joy at the loss or gain of a boyfriend or a girlfriend. The interaction may be quite intense at different periods of the day (immediately before and after school) or the weekend (Friday and Saturday afternoon and evening) and if one or another member is lax in their participation, this can be interpreted as a breach in their responsibilities with respect to the group (Rivère and Licoppe 2005, p. 14). The participants characterized here are generally younger than those discussed above when thinking about the conversational form of interaction. Indeed, there are generational differences to be observed here.

The emphasis in connected presence is on the strong bonds of interaction.[31] The lore of the group is developed, commented on, and crystallized.

The power relations are defined, reaffirmed, and developed through a seeming perpetual stream of more—but often less—important communications. A stable sense of the group is developed through the mutual engagement of the various persons in the social network. The sum of this interaction is the development and maintenance of a collective ideological perspective.

The mobile phone is an enabling technology for connected presence. Whenever we feel the need or the desire to interact with our intimates, we can act on the urge then and there. We do not need to collect the themes we can imagine talking about with others and saving them until our next meeting, or, as in the more conversational mode of telephonic interaction, until the following Saturday, when we will have our regular call. The interaction can take place at the drop of a hat. We can call or we can send a message spontaneously. Indeed, some teens practice interaction of this sort far into the night.[32]

Thus, the mobile phone structures the flux of interactions, the flow of information, and the sense of belongingness in the group. It is precisely the omnipresence of the mobile telephone that facilitates this intense form of interaction:

With regard to mediated interactions, we can, for example, consider that the fact of maintaining this connected presence, ratified by the interlocutor, allows for a lesser formality of mediated interaction: it becomes less necessary to reassert the formal and institutional aspects of the frame of interaction at each call if one is feeling connected to the other person through a continuous flow of small communicative acts. (Licoppe 2004, p. 154)

To be sure, not all users are drawn into this kind of interaction. Many mobile phone users make or receive a call only occasionally. Older users who favor the more staid form of conversational interaction are perhaps unaffected. However, as the research cited above indicates, there are those whose attention is drawn to the group through this kind of interaction. It is in the peer group that teens and young adults are often able to construct a social network characterized by the excitement of "being together." It is here that mobile communication provides the elixir of communication. The research cited above indicates that the use of mobile communication is a phenomenon of the small group. The device allows the members to exchange information and confirm the group's internal linkage in a far more "connected" form. It allows for the exchanging of jokes and gossip. It allows for the development of argot in texting and it provides the stage upon which Romeo and Juliet wannabes can work out the crossing of their stars.

Conclusion

The mobile phone has become intertwined in our daily activities in a remarkably short time. In Norway, for example, there were few teens that owned one in the mid 1990s. Assisted by post-paid subscriptions and by access to relatively inexpensive handsets, the situation quickly shifted such that by 2005 there were no 15-year-old teens that did not have one.[33] This transition is not simply in terms of owning a device that perhaps sits on the shelf and endlessly blinks 12:00. The material here shows that the mobile telephone has been integrated into daily life both in overt situations and also covertly, as with the students who manage their social lives in mid-lecture. All of this is so well integrated into the daily flow of events that it is, in the literal sense, not remarkable. This, along with the low technical threshold associated with communicating, means that the peer group and the family have a channel through which they can easily contact one another, keep each other abreast of their activities, and unburden themselves of their various thoughts and insights. In addition, we use the device to engage in ritual interactions that help us to feel secure in the ongoing maintenance of our social world. While this might ruffle the feathers of others we meet on the street, it also supports in-group interaction.

Voice interaction and texting have resulted in tighter integration of the group. To a degree, co-presence and mediated interaction are not seen as absolute categories. Ito describes how a straggler announces his or her "presence" at a meeting of the gang with a message saying "I will be there in five minutes." This is enough to allow the others in the group to orient themselves in terms of their mutual interaction instead of wasting time on wondering as to the status of the laggard. Following from Licoppe's notion of connected presence, it is also likely co-present members of the group will know that the latecomer has had a tough day at work or has argued with his or her girlfriend or boyfriend. Upon their arrival, these topics may be common knowledge in the group, and aside from some of the juicy details the group is already in the process of forming their opinion of the situation and placing the recent events into the broader dogma of the group (bosses are overbearing, boyfriends or girlfriends are not to be trusted, etc.).

The treatment of the latecomer also brings us back to the notion of ritual interaction. The sharing of the latest gossip and perhaps the laughter at the expense of the late arrival may have been initiated via a mobile connection and developed in a co-present context. In this way we see that the phone does not just give access to one another; it enhances in-group interaction and thus facilitates the types of rituals discussed above. It makes clear who is included in the in-group and in the out-group. It can help in setting a mood, and it facilitates the mutual engagement of the group. These events can make explicit the internal structure of the group. These interactions can provide insight into when the internal status of the group is aligned and in other situations when the boundaries are being tested. The individuals have their ways of engaging one another, their inside jokes, their sense of the power structure of the group, and their lines of intrigue. As we will see in the next chapter, these interactions can be seen in terms of the group's sense of local ideology. In short, the mediated interaction of the group can make explicit the ongoing project of producing group identity and cohesion.

10

The Recalibration of Social Cohesion

In chapter 2, I posed the question of whether society was becoming more cohesive or crumbling into a universe of individuals. After setting that somewhat dismal stage, I began to examine the development of thought surrounding the use of ritual and its role in the cultivation of social cohesion. From Durkheim's form of "authored" rituals to Goffman's and Collins's sense of do-it-yourself rituals, I traced how this kind of interaction permeates our daily interaction.

In chapters 7 and 8, I examined how the mobile telephone affects co-present interaction and mediated interaction. The general sense was that the device does not always fit into the flow of co-present interaction, but that when thinking of mediated interaction within the group, mobile communication facilitates the development of cohesion through the use of various ritual devices (argot, greeting forms, humor, gossip) and by playing cupid for lovers. The use of Goffman's interaction rituals provides a format through which the in-group works out a sense of common identity and a sense of unity. It is in this forum, both when co-located and when they interact via mobile communication, that we forge the links of friendship and solidarity that can be seen as social cohesion.

In chapter 9, I examined research showing that mobile communication seems to have found its niche in the intimate sphere (friends, lovers, family). I suggest that this is due to the ritual interaction described in chapter 8 and, as Licoppe suggests, to continual voice and text-based contact, carried out in short bursts of interaction through the day. In short, mobile communication is becoming a part of small-group interaction.

In this chapter, I will examine one final way in which cohesion is established: through the use of group ideology. I will look at how we use the ritual form to develop a locally minted ideology. It is from this that the group draws its sense of uniqueness, negotiates power relations, and marks boundaries.

After considering the ideological dimensions of the group, I will then return to an issue considered in chapter 2: the issue of social cohesion in society. There is much discussion as to the general direction of society when confronted with the new digital technologies (Ling 2004d). There is the sense by some that these technologies are causing a drift toward individualism (Beck and Beck-Gernsheim 2002; Bugeja 2005; Kraut et al. 1998; McPherson et al. 2006; Nie et al. 2002; Putnam 2000; Wolak 2003). On the other hand, some suggest that they are supporting new forms of sociation (Boase et al. 2006; Katz and Rice 2002; Kavanaugh and Patterson 2001; McIntosh and Harwood 2002; Turkle 1995).

When looked at through the lens of mobile telephony, the material here seems to point at a kind of middle solution. We have seen that there is some fraying in civil society. While far from being conclusive, the material in chapter 7 indicates that when, for example, one is given the choice between conversing with another marginally known person at a bus stop and sending a text message to a paramour, there is little doubt as to the outcome. More often than not, we choose the familiar person rather than the marginally known. At the same time, the analysis in chapters 8 and 9 show that the small group seems to be benefiting from the use of mobile communication. Thus, we see that there is a tightening in the individual's social network that augurs against those who are marginally known to us and in favor of those who are familiar. That is, we are perhaps seeing the development of bounded solidarity. That is, we are seeing the potential for the clique to be so focused on its own interactions that the so-called weak-link connections are neglected.

What is the correct level of cohesion? This is not a simple too-much-or-too-little question. The issue here is balance between the strong ties of the core group and the weak ties toward other points in the social compass.

Bounded Solidarity: The Development and Maintenance of Local Ideologies

Local ideologies are an important aspect of bounded solidarity. A local ideology has three parts. First it must have some underpinning structure that outlines the general orientation for the group. In addition to this, a local ideology needs to be refreshed from time to time with new examples showing why the ideology is correct. Finally, there needs to be the willingness to enforce a discipline that guards the basic tenets of the ideology. These elements can be seen in the transcript of an interview conducted in Norway in 1997. This was a much earlier point in the adoption process for mobile phones, and the family being interviewed (consisting of a mother, father, and a 15-year-old son) was fighting a rear-guard action against the use of mobile telephony:

Interviewer[1]:   What about mobile phones? Do any of you have . . .

Martin (son):   None of us has a mobile phone. We are pretty much against mobile phones.

Henrik (father):   We are against [mobile phones] because I see as a photographer, I recently had, I photograph weddings. I recently had a couple that I photographed last Saturday. The groom was continually using the mobile telephone while I was taking pictures. I said "either you put that down or I am going to stop taking pictures." And so he put it away and I said "Now you turn on the mobile answering or I will stop taking pictures." And I also had a situation where it rang in the church during the ceremony and he begin to talk. I don't have a mobile telephone and I don't want one because I have seen too much negative use of them. People use them regardless of if they are on the train or driving their car, so people almost collide and they sit on the train and talk. They use it too much. It is not necessary and I see teens walking around and talking . . .

Martin: And in school . . .

Elise (Mother):   They are very dangerous. I have seen several near-collisions with mobile telephones where they are talking on the mobile telephone and have almost collided.

Henrik: I have seen a lot of near-collisions with the mobile phone, where they are talking on the mobile phone and have almost crashed.

Martin: There are some in my class, you know. There are quite a few that are like snobs and have real expensive clothes and such and a lot of them have like a mobile telephone to show that they are rich. . . .

Here we can see the functioning of a local ideology. There is the underpinning structure ("We are pretty much against mobile phones"). This local ideology was so well formulated that the interviewer was not even able to complete the question before Martin interrupted with the "party line." In addition, there is the cultivation of the ideology and its subsidiary points by using various supporting arguments enhanced by fresh examples of why mobile phones are not a good thing.[2] In this example, the family has four secondary justifications for not liking mobile phones. They outline how the mobile phone is a disturbance, dangerous when one is driving, and a snobbish status symbol,[3] and that there is no need to be perpetually available. These elements are also illustrated by fresh examples of their undesirability. The father recites the frustrating situation with the groom when he was trying to photograph the couple, the mother reports having seen "several near-collisions," and the son tells about how the device is flaunted in his school.

Somewhat later in the discussion, Elise opens up the consideration of situations where a mobile phone might be acceptable, and Henrik muses as to whether it might be advantageous to be available to his customers via a pager, a technology that was still available at that point. This affords the chance to reexamine their general attitude toward mobile communication and eventually the need to restate the basic tenets of the ideology:

Elise: I have been out with students [she is a teacher at a nursing school] who have, ahh, that their mobile phone has rung. But it is usually that they have small children that are sick so that in a way it is an emergency telephone, that is OK. But I can't say that I accept it.

Henrik: That is another thing. I could imagine having one of those, what is it called, a pager. I have thought about that. So that people can get in touch with me when I am out driving. That way you don't have to talk, just a telephone number comes. I have wondered about that because then people can get a hold of me that way. When I am out taking pictures and somebody else that wants to contact me, if I can come to here or there. And the family can just send me a code if they want me to call home. That is OK.

Elise: But I have been against that a little also because I just think that sometimes. You don't always have to be available . . . I don't think that it is correct that we can always be available. When you drive a car you need to be free from the stress of having a telephone. If you are out on a walk you don't need the stress with both pagers and things like that. I think that it is enough with an answering machine.

The local ideology is well established but not completely set. Elise expresses certain openness to the local ideology concerning mobile communication. Her assertion is that mobile communication may have its place in emergency situations, or at least in the case of a child's sickness.

Henrik then starts to examine the potential for better coordination provided by mobile communication. However, after this short excursion to the other side of the ideological divide, Elise comes back to the more basic tenets of the ideology and thus ends the discussion. Elise takes her role as the defender of the ideology.

In the illustration here we can see how a small group (in this case, family members who are discussing an issue face to face) develops and maintains a local ideology. Local ideologies can also be developed and supplemented via mediated interaction. The case cited here has to do with opposition to mobile phones, but local ideologies are easily applied to other realms. Mobile communication can be a tool through which these local ideologies are developed, strengthened, and maintained. For example, within the teen peer group the ideology can be something as simple as the general idea that fathers are jerks. As with the family described above, if this was the only aspect of the ideology it would soon be dead. In order for an ideology to be living, it must have a well-annunciated base (for example, that fathers are jerks). In addition, it relies on a constant stream of new instances and illustrations (Dad wouldn't let me go to the movies because the relatives are coming over. Gee, he is a jerk!). In this way, the base ideology is developed and embroidered over time. The situations in which fathers are jerks may need to be better defined (fathers are jerks, unless Dad pays for the new dress I want.) and the general ideology may have to be defended (I know your father bought you a new dress, but remember when he wouldn't let you go to the movies? Fathers are jerks!) The ideology may mutate (Dad is a jerk when we are at home, but he is OK when we are on vacation) or it broaden (Parents are jerks!). In any case, an ideology must have a core element and a stream of examples confirming its validity.

Just as with the example of the family examined above, there needs to be a constant flow of events that in some way illustrate the validity of the ideology. These need to become part of the lore of the group. Thus, they need to be communicated and shared. The mobile phone is an ideal device in this way. It allows the person-to-person communication of these messages.

∎

The notion of small groups having unique ideological perspectives is not new. Berger and Kellner have discussed the idea of the nomos. They look at this as being the opposite of Durkheim's anomie: "Just as the individual's deprivation of relationship with his significant others will plunge him into anomie, so their continued presence will sustain for him that nomos by which he can feel at home in the world." (1964, p. 7)

Whereas anomie indicates that an individual does not have any social moorings, nomos would suggest that the individual is bounded by various socially constructed devices. In particular, Berger and Kellner apply the development of nomos to the situation of marriage and the construction of a collective sense of identity.[4]

Another approach to understanding the internal dynamics of small groups is provided by Gary Fine (1979). Fine develops the idea of idioculture: "a system of knowledge, beliefs, behaviors, and customs shared by members of an interacting group to which members can refer and employ as the basis of further interaction. Members recognize that they share experiences in common and these experiences can be referred to with the expectation that they will be understood by other members, and further can be employed to construct a social reality. The term, stressing the localized nature of culture, implies that it need not be part of a demographically distinct subgroup, but rather that it is a particularistic development of any group in the society." (ibid., p. 734) Fine examines how this locally produced sense of common identity is the result of group interaction and how when it is once formed, it in turn prefigures the interaction of group and their process of understanding various contingencies. If applied to the mobile-phone-averse family described above, the establishment of the basic notion of not liking mobile phones gets worked out in various ways of understanding others' use of the devices (pompous, dangerous, stressed). The idioculture can be changed and adjusted as various exigencies present themselves. Nonetheless, once the ideological baggage is in place, it provides a general line of action. The family described above cannot glibly go out one Saturday and buy mobile phones for Mom, Dad, and Junior. They would have had to work through some special process of acceptance or necessity in order to get around their pre-constructed ideological elements. In the case of mobile telephony, it is indeed often the case that a previously held opposition to the device is set aside when confronted with a local crisis such as the illness of an elderly parent.

On the ideological plain, the person who has most directly examined this issue is Kenneth Gergen (2003, 2008). Gergen examines the role of mobile communication with respect to the broader political drift of society. He suggests most interactive media contribute to wide-ranging but perhaps superficial relationships. The mobile phone has the opposite characteristics. Mobile communication facilitates in-group interaction[5] to the degree that it hampers individuals' ability to engage in broader political interaction. Rather, the individuals form self-sustained groups that are inwardly focused:

> . . . communication technology has hastened the erosion of civil culture, mobile communication in particular has played a critical role in bringing about transformation. Essentially we are witnessing a shift from civil society to monadic clusters of close relationships. These "floating worlds" of communication (Gergen 2003) enable such groups to remain in virtually continuous contact. (Gergen 2008)

Gergen speculates that the concern for a broader sense of society is replaced by an ethic that maintains the group. This ethic or ideology can include what is safe or risky humor/gossip and the ideology can cover what argot is appropriate in texting as well as the type and form of romantic interaction. The ritual interactions of the group described here provide the mechanism with which those within the sphere of the group integrate, or perhaps over-integrate themselves.

Where Licoppe perhaps sees "connected presence" as a revitalization of the primary group, Gergen sees it as a retreat from the public sphere, a way to buffer ourselves away from the problems of the broader society.[6] In these groupings, the main concern is the mundane issues of the group itself. Engagements in the broader social order are of concern only to the degree that they directly influence the flux of the monadic group. The mobile phone, however, mostly functions as a kind of tether between the individual and the group. Consider this passage from Gergen 2008:

> The mobile phone brings small enclaves into continuous co-presence, so do actions that might otherwise be attributed to the autonomous self become embedded in dialogue. Private thought and public deliberation converge; isolated emotionality is replaced by emotional sharing (Rivère 2002). We increasingly locate practical reasoning in relationships as opposed to independent minds (Katz and Aakhus 2002a). Personal significance is acquired through one's place within the network. The mobile phone functions symbolically, then, as an umbilical cord through which one draws vital nurturance from a larger, nurturing and protective force. It is not one's individual thought and personal desire that is now central to the democratic process but relational interchange.

Personal significance is acquired through one's place within the network. The mobile phone functions symbolically, then, as a link through which we draw sustenance from a larger group.

This line of thought leads to the notion of what we might call "Bounded solidarity." That is, individuals exist in a situation where there is a kind of reciprocal affirmation. In this configuration, group members exchange and share, re-share and pool small variations of each other's ideas (Gergen 2008). The system of interaction is relatively closed to new insights. Closed networks do not enhance information flow so much as they create an echo that reinforces existing predispositions. Information shared in this way is not properly discounted. It can result in an erroneous sense of certainty. It can enhance the positions of the group at the expense of reliability. In addition, the amplification of predispositions leads to internal trust where it may not be justified (Burt 2001a). We can speculate that in such a closed social network there is a high degree of common insight into others' thoughts but there is no fresh input. The ritual interaction within the group along with the production of a local ideology tightens the interaction within the group to the degree that it is largely the internal dynamics that are primary. To the degree that this situation obtains, there is an over-reliance on the internal strong ties and little of Granovetter's weak ties (Granovetter1973; see also Burt 2000, 2001b).

According to Portes, there are several issues that arise when the internal ties are over-configured. These include the exclusion of outsiders, excessive claims on group membership, restrictions on individual freedom, and downward leveling of norms (Portes 1998). In addition, when the internal strong ties are supported by a local ideology, the dynamic of the group can become resistant to other influences.

These issues can be played out in teen peer groups. The members can shun others based on fine-grained nuances (Lynne 2000) and command the attention of members where normally there would be more openness to other more distant, and more varied individuals. It is in situations like this that the group does not exploit other types of social resources that are usually made available through weak ties. If the group is to internally focused while they are in complete agreement about the correct style of clothing and the best music, their internal focus means that they do not pick up the influences of others. The clique has the ability to reserve the time and energy of the individual. The group dynamics become a kind of internal taxation where the expectations of reciprocity are so heavy that the individuals are limited in their ability to take individual initiative.

ICT and the Expansion of Social Horizons

The line of thought has taken a rather gloomy turn. The argument presented above speculates that the mobile phone is influencing sociation too well. It is closing us off, focusing us on our mobile interlocutors to the degree that we do not smell the social flowers around us. The use of ritual interaction has allowed the individual to cultivate their peer associations remotely. This has tipped the balance in the direction of the peer group at the expense of the broader social scene. This is, however, an oversimplification. It is a tipping of the balance, but perhaps it is only that (Harper 2003). It is not upsetting the balance. Just as Gergen can speculate that mobile communication is pushing in the direction of reduced engagement in the political sphere, it can also facilitate the opposite. There is the ability to work out new relationships (Ellwood-Clayton 2005; Prøitz 2004) and to organize broad political interactions (Castells et al. 2004).

While the processes describe a change in the internal dynamics of the peer group or family, there has only been superficial analysis of how mobile communication impacts on, for example political engagement. In Norway, there is, in fact, weak evidence that points in the opposite direction. Date from a survey of more than 11,000 teens[7] indicate that those who sent more text messages were also likely to have been more active in student politics.[8] They were more likely to have participated in a political party.[9] The same is evident when looking at using the voice function of mobile communication and participation in student politics[10] and having been active in a formal political party.[11]

∎

International research shows that there is multiplexity in our interaction with others (Haythornthwaite 2001). Email, social networking via Facebook, and other forms of online interaction supplement co-present and telephonic contact. In addition, they supplement our activity in voluntary organizations (Quan-Haase and Wellman 2002; Kavanaugh and Reese 2005; Kavanaugh and Patterson 2001). A somewhat similar picture emerges from data on instant messaging. Rather than only engaging with the same tried-and-true friends, Shiu and Lenhart (2004, p. 13) found, teens were quite active in terms of adding new names to their "buddy lists":

Gen Y instant messengers manage their buddy lists much more actively than other IM users. Fourteen percent of Gen Y-ers add someone new to their buddy list at least once a week, an activity that 2 percent or less of every other generation engages in. Gen Y-ers are also more likely to engage in several separate IM conversations at the same time. Nearly a third (29 percent) do so everyday—a percentage that is more than double the next highest proportion.

In focus groups in Oslo, teens have reported that instant messaging and social networking are seen as more open. Whereas mobile communication is reserved for the close peer group, IM lists and sets of Facebook or MySpace friends are usually broader. This is due in part to the fact that when you log onto IM, other more marginally known contacts can take the opportunity to broach contact (Ling and Baron 2007).[12] In addition, there is the normal pattern of co-present interaction that while it is being affected by the encroachment of information and communication technologies, it is still a rich source of stimulation. All told, these channels allow for the mechanisms described by Travers and Milgram in their "small world" model (1969). As the number of links available to an individual increases, there is a more open flow of information (Barabási 2002). Technologies such as IM, social networking on the net, and the normal flux of life mean that we often come into contact with many different people. Thus, the hermetic tendencies of mobile communication would have to be quite powerful were they to isolate the clique.

Another, perhaps slower moving dynamic is that the bounded groups are not immutable. Transitions are a part of adolescence (Ling and Helmersen 2000). Some transitions are occasioned by changes in taste and by inter-group frictions that may draw one person in at the expense of another. Others are occasioned by life-cycle issues of proceeding from transitions (e.g., from junior high to high school to college or a career). Others are seen as young adults move from one wave of fashion to another and from one partner to another.

Thus, while there is a movement toward greater interaction in the small group, this is not the only process taking place. While mobile telephony draws the focus to the intimate sphere, it and other information and communication technologies (ICTs) also serve to open up the horizon. The analyses show that there is openness to external influences, at least among teens in their use of ICTs. On the one hand, those who use mobile communication are engaged in the broader political drift of events and second, that there are alternative forums through which these links are being realized. On the other hand, the technology, and in particular the mobile phone is bringing with it a new form for direct interpersonal interaction. Indeed, as Putnam suggests we have perhaps more to fear from the individualizing influences of TV and the suburbs than we do from the monadic propensity of the mobile phone. Drawing this back to the speculations of Gergen, it is not clear whether he is being pessimistic or prescient.

Individualization, Communalization, or What?

Mobile communication is enhancing interaction within the small group. At the same time, the data here suggests that this is not necessarily displacing interaction with the broader social sphere.

So what is it? Are we becoming more individualized, or less so? A simple approach to this would suggest that there is a linear movement away from Tönnies's organic, reciprocal, tradition-oriented Gemeinschaft to the more "contractual" and rational Gesellschaft society. As Lash (1994) noted, we are currently at some point on this continuum. We don't have the structure of village, church, and family, but we have not arrived at the ultimate Gesellschaft position of complete individual agency. According to Lash, the current situation is characterized by varying degrees of quasi-Gesellschaft. That is, our interactions with the broader world are often moderated via the influences of trade unions, the welfare state, and various formal entities. Lasch notes that we do not arrive at a state of Gesellschaft until there are no mediating institutions between the individual and society (1994, p. 114). The prime motivating factor was the process of individualization that motivated the whole process.

The tension here draws on another part of Durkheim's theoretical legacy: the interaction of mechanical and organic solidarity (Durkheim 1997). In the small group, we find the legacy of mechanical solidarity, that is group conformity, local ideology, group lore, and group boundedness (Collins 2004a, p. 348; Scheff 1990, p. 22). The counterpoint to mechanical solidarity is organic solidarity, which Durkheim suggests is the general movement toward more abstract and formalistic membership in society (Cladis 2005; Rawls 2003). This form of solidarity is, according to Durkheim, a characteristic of complex and differentiated society. In the Durkheimian sense, individualization is not a process that the individual carries out; it is a more general process that works itself out in the life histories of individuals.

The material presented in this book paints a somewhat different picture. Although the tendency in society has been toward individualism (or, in the Durkheimian sense, toward organic solidarity), there are also counter-tendencies. The material in chapter 8, in chapter 9, and earlier in the present chapter points to what might be a small revival of the familiar sphere—Durkheim's mechanical solidarity on a local level (that is, a bounded or limited solidarity).[13] The material here has examined how in a world that is perhaps characterized as moving toward organic solidarity, the mobile phone opens a space for a neo-mechanical solidarity.

The small group of friends is contacting one another with more frequency and the family is better able to monitor one another's needs. We are in continual, if brief and topical contact. As Katz et al. (2004) suggest, the ritual mechanisms with which to develop and maintain local ideologies are alive and well. Billions of text messages fly back and forth telling of friends' plans and their random thoughts. We receive phatic interactions seemingly at the drop of a hat.

The assertion here is that these forms of ritual interaction are helping the peer group and the family to form cohesive bonds in ways that were not available in the recent past. Teenage lovers are exchanging endearments, parents are coordinating daily life, and juvenile delinquents are planning their next action. That is, there has been a recalibration of how social cohesion is being worked out in society. The material presented here indicates that, while there is a tightening of the local group, it is not necessarily at the expense of involvement in the broader social flux of activity. We are picking up on the ideas and trends being discussed in other locations and by other groups. We have the ability to dig up obscure facts on marginal activities and bring these into our daily interactions. Thus, on the one hand, the familiar is becoming more amplified and on the other, the range of interaction is being extended.

The simple progression from Gemeinschaft to Gesellschaft is difficult to fit into this typology. There does not seem to be the drift toward individualization as described by Lash. Instead of being a quasi-Gesellschaft of Lash, there is what Merton called the pseudo-Gemeinschaft (1968; see also Katz et al. 2004). Rather than a quasi-corporate stop on the way to full individualism, perhaps we are dealing with the legacy of earlier, more communal forms of interaction. These are not as stiff as those defined by the church, the family, and tradition, but they still retain the human touch of social interaction that is a buffer against the other individualistic winds blowing in society. With mobile communication and other forms of net-based interaction, there is a strengthening of the in-group dynamics. At the same time, the cultural capital of the group can be spiced up with other bits of social flotsam and jetsam.

The low threshold for communication in our primary groups makes them more robust and also allows them to develop their own sense of self-identity. The names and numbers of all of the gang are readily available, thus facilitating the flow of chatter, be it instrumental or expressive. Shifting between co-present and mediated interaction allows us to develop and elaborate our jokes, gossip, and argot. The mutual sense of mood and common insight into the flux of events is shared both face to face and via mediated interaction. We engender mutual recognition of a common mood within our group. Teens are able to understand each other's angst regarding an upcoming party or the potential that their team will win an upcoming tournament. Further, there is not the need to wait until the next "flesh meet" in order to exchange information or comments. We can do it spontaneously, whenever and wherever the muse strikes. In short, Simmel's (1949) "impulse to sociability" is holding its own against the striving toward individualism (Beck et al. 1994)

Mobile communication has changed the way we interrelate. It has also changed our expectations of social interaction. My meeting with the plumber (chapter 1) illustrates much of this. He was clearly more absorbed in his mobile phone conversation than in the etiquette of meeting new customers. He was a nice enough fellow, and he did me the favor of providing an ingress and an exit for this book.

The plumber's engrossment in the phone conversation was, in some ways, to be expected. It was probably was his boss or his girlfriend on the other end of the line—someone to whom he had to give his attention. In the balancing act between his telephonic interaction and the need to deal with just another job, I was put into my proper place. In a somewhat bemused (or perhaps confused) way, I yielded to his behavior. That's OK; it wasn't the first time I found myself in such a situation, and it won't be the last.

Notes

Chapter 1

1. Social cohesion is obvious a complex issue. In this book, I use the term in the sense of individuals wishing to join and maintain a group. See, e.g., Friedkin 2004.

2. These forms of interaction are starting to blend. It is now possible to receive email via mobile communications devices (the Japanese mobile phone, commonly called a keitai, the BlackBerry, and most "smart phones").

3. Putnam has been criticized for his failure to examine the gendering of social capital and his inability to apply his findings to other countries. See chapter 2.

4. This approach is not new to communications studies. Drawing on Durkheim, Carey (1992) talks about a ritual view of communication.

5. According to Collins, a ritual is "a mechanism of mutually focused emotion and attention producing a momentarily shared reality which thereby generates solidarity and symbols of group membership" (2004a, p. 7). It is not necessarily a repeated or an obsessive behavior.

6. SMS (Short Message System) is the usual form of textual interaction between mobile telephones. However, in Japan, with the iMode system, the messages are technically somewhat closer to email. Thus, a generic term for textual interaction between mobile communication devices is 'texting'. This may change as instant messaging functionality becomes a part of mobile telephony.

7. Park (2005) uses a milder form of this definition in his research on addiction to mobile telephony.

8. See also Franklin et al. 2000.

9. Durkheim uses the word 'effervescence' to describe a more general stimulation of energies wherein "people live differently than in normal times" (1995, p. 213). It is used to describe the excitement associated with ritual interaction.

10. I am not, however, categorical in saying that these bonds cannot be forged in other contexts. In the mid 1990s there was an active line of research in the development and elaboration of "internet" relationships (Lea and Spears 1995; Ling 2000). The general sense was that while these are quite rare, the relationships

that develop on the net can form the basis for completely traditional co-present relationships.

11. The connection between driving and telephony has shown itself to be a lasting issue (Jessop 2006).

12. The development in Australia came at about the same time (Goggin 2006).

13. I do not capitalize 'internet' or 'web'.

14. Obviously not everybody has several telephones. This statistic is the result of subscriptions being assigned to functions (ambulances, various machines that can "call" for service. etc.). These include soda machines and other remotely placed devices that require periodic maintenance). In addition, there are often many old unused pre-paid subscriptions that are on the books of the telecom operators. For a further discussion of this, see Ling 2004b.

15. In China, there were 30 telephones for every 100 persons in 2005. The US had 68 mobile phone subscriptions per 100 persons, Russia had 83, Japan had 73, and Brazil had 46.

16. It is not clear from this material whether messaging via CDMA or iMode was included in the count. Also, the standard of reporting for these types of events varies dramatically from country to country.

17. The data here are derived from a nationally representative sample of 1,000 users in a survey carried out by Telenor in June 2006.

18. In addition to the attractive force of rituals, groups also find cohesion when confronted with the fear of other alien groups.

19. This encouragement came at the first of several seminars hosted by James Katz at Rutgers University. See also Schegloff 1998.

20. Okabe and Ito (2005) use a similar form of observation in their work on mobile telephony in Japan.

21. Indeed, both Goffman and Bourdieu used cameras at various points in their field work (Witkin 2006).

Chapter 2

1. For an examination of these issues in the context of communications, see Katz et al. 2004.

2. There are broad similarities between social capital and the discussions of communities of practice (Duguid 2005). While there are some intramural squabbles on some of the finer points, the two approaches generally look at the ways in which social interaction is a unique phenomenon as opposed to the more individually oriented social sciences of psychology and economics. The community of practice literature has contributed the concept of tacit knowledge to the discussion and this is a fortunate contribution because it addresses our focus on the existence of uncodified and perhaps uncodifiable knowledge.

3. Bourdieu's concept of cultural capital that is often translated to mean taste or sophistication. He notes that cultural capital is "convertible, on certain con-

ditions, into economic capital and may be institutionalized in the form of educational qualifications" (1995, p. 243). He talks about this as being an accumulation of poise or elegance and, like economic capital it is a characteristic of the individual. Like economic capital the individual can also "invest" in cultural capital by, for example taking classes on art history or wine tasting. Thus, it is not only the possession of the appropriate artifacts, but it also can include understanding their significance relative to other similar items. This kind of capital is, perhaps more varied than economic capital in that we can become connoisseur of various types of cultural items. We can become an aficionado of Van Gogh, handmade fountain pens, or Thelonious Monk. Bourdieu suggests that cultural capital is the realm of academics. Indeed, he sees academic qualification as a "certificate of cultural competence" (1985, p. 248) Where economic and cultural capital are characteristics of the individual, social capital is a characteristic of the group. While the individual may enjoy the advantages or strain under the requirements of social capital, it is not just a trait of one person. Rather it is a characteristic of that person in the context of their social group or network. Bourdieu says that social capital is "made up of social obligations ("connections")" (ibid., p. 243). He also notes that "the aggregate of the actual or potential resources which are linked to possession of a durable network of more or less institutionalized relationships of mutual acquaintance and recognition—or in other words, to membership in a group—which provides each of its members with the backing of the collectively, a 'credential' which entitles them to credit, in the various senses of the word" (ibid., pp. 248–249).

4. Indeed, John et al. (2003) have found that trust and not social networking are central in terms of social capital.

5. In the same study, Coleman also examines the social capital of Korean dissident students and the functioning of neighborhoods with reference to children's safety in Jerusalem and in Detroit.

6. Hardin (1968) poses a similar question in his essay on the tragedy of the commons.

7. Putnam cites work showing that lack of so-called weak ties contributes to the preservation of impoverishment since one literally does not receive information about possible job opportunities.

8. Others question whether there is even a need to study the collective. Duguid (2005), who is interested in the role of tacit (i.e., unspoken, uncodified, and perhaps uncodifiable) knowledge that often constitutes the core of social capital, cites Simon, who feels that there is no such thing as tacit knowledge.

Chapter 3

1. Durkheim's *Suicide* (1951/1897) was, however, heralded, as it showed the insight to be gained by careful social analysis. In that book, Durkheim took the dramatic event of suicide out of the purely psychological realm and examined it in social terms. His distinctions between, egoistic, anomic, and altruistic based on religious culture or social structure secured his place in the panoply of great

sociologists. He was able to show how the social order has a hand even in a situation where the individual is seemingly engaged in the most personal of all events.

2. According to Duncan (1970), Durkheim asserts that religion is at the core of the social order. This assertion is parried here by suggesting that it is ritual processes.

3. For a broader discussion of the re-application of Durkheim at the micro (or perhaps meso) level, see Rawls 1996.

4. See Collins 2005, p. 126.

5. This idea of social density is seen in Durkheim's discussion of suicide (1951), in which he examines how the structure of the society—and in particular the religious practices of a society relate to the prevalence and form of suicide. In social regimes where the individual is more tightly integrated into a social structure, there was less suicide. Thus, as Durkheim could show, there was less suicide among Jews and Catholics than among Protestants. The idea was that the price Protestants paid for allowing for more individualism in their ideology was reduced social ties and the lack of a social structure that would maintain them when times got tough. This line of thought has also been applied to juvenile deviance (Elliott and Ageton 1979).

6. It is clearly easy for individuals to become the focus of the ritual obeisance. We only need to think of the cults of personality that have arisen and manipulated by unscrupulous leaders such as Ceausescu and Amin. The line here is unclear, however. Leaders in some situations become prisoners to their ritual role. They are trussed up by the role more than they may have expected at the outset, and in a sense they are caged in a lifelong captivity. (Examples include some US Supreme Court judges, some "royals," and some popes.) In this case there is a person who is not involved in the abuses of ritual position but who has been encased in a ritual position.

7. Born in 1858 in Epinal, in the northwest of France, Durkheim broke the rabbinical tradition of his family and eventually broke with Judaism completely. In 1870, France was defeated by the Germans. It has been speculated that the experience of Prussian anti-Semitism during their occupation of the Alsace-Lorraine region in the 1870s provided Durkheim with direct insight into the second element of social cohesion, namely opposition between groups. Later, after the collapse of the Commune of Paris and the events following that there was a period of nationalism in France lasting into the mid 1880s that also included anti-Semitic elements. After distinguishing himself locally as a student, he left Epinal for Paris in order to prepare for admission into the Ecole Normale Supérieure, where, after some setbacks, he was admitted in 1879. As a student it seems that Durkheim was active both intellectually and politically in the discussions of the day. This included a movement away from the more traditional studies of language and rhetoric and more of a focus on empirical and scientific analysis. In 1882 Durkheim passed an exam that allowed him to teach in secondary school. On the basis of his work regarding German Philosophy, he was appointed Chargé d'un Cours de Science Sociale et de Pédagogie at Bordeaux in 1887. He remained there until

1902, when he returned to Paris. In Bordeaux he had a primary responsibility for lecturing on pedagogy, as that was the main focus of his appointment. He also lectured on areas of social science. According to Jones, he developed many of his idea on issues such as social solidarity, family and kinship, incest, totemism, suicide, crime, religion, socialism, and law in these lectures. It is also during this period that he wrote three of his major contributions to sociology: *De la division du travail social* (1893), *Le Suicide* (1897), and *Les Règles de la méthode sociologique* (1895). Also during the latter part of this period he founded *Année sociologique*, the first social science journal in France. In 1902, Durkheim returned to Paris to the Faculty of Letters at the Sorbonne, where he assumed a chair that was eventually named the chair of Science of Education and Sociology. With the start of World War I, Durkheim's life took a new turn. After the invasion of France, Durkheim was active in the publication of documents and studies on the war that were intended to help sway other countries in their support of France. However, again suffering from his situation as a Jew from Lorraine with a German last name, he suffered from a certain mistrust of a vulgar sort (Jones 1986). The most severe result, however, was the death of his son André in the war. Durkheim died of a stroke on November 15, 1917 at the age of 59.

8. Couldry discusses this in the context of large-scale media events where the individual is, vicariously, transported into a common situation. Obviously, the use of point-to-point meditation as considered here requires that we scale scaled back to the small group and the interpersonal level when dealing with mobile communication.

9. The closest Goffman comes to discussing totems is his discussion of "identity kits" by patients in a mental ward (1961, pp. 20–22).

10. Mary Douglas (1966, p. 80) suggests that the work of Radcliff-Brown and Malinowski did the service of tearing down the barrier between religious and secular ritual.

11. One point here is that it is tempting to look only at the polar examples. To think only in terms of the exclusively mediated vs. the exclusively co-present form of interaction. The situation today is not so clear cut. It is becoming increasingly difficult to find the pure cases, and indeed it is becoming less and less interesting. Rather, we need to think about social cohesion that relies on more or less mediated interaction.

Chapter 4

1. Bourdieu makes some of the same assertions in the sense that he finds in "insignificant" details he traces to the broader culture. He describes how, for example, injunctions regarding which hand to use when eating or how one should sit in a chair describe broader social attitudes (Bourdieu 2002; see also Couldry 2003).

2. In a 1980 interview, Goffman said: "My main influences were [Lloyd] Warner and [A. R.] Radcliff-Brown, [Emile] Durkheim, and [Everett] Hughes. Maybe [Max] Weber also." (Goffman and Verhoeven 1993, p. 321)

3. Another of Durkheim's "intellectual children" is his influence on the structural-functionalism of Parsons (Collins 2005).

4. One of my teachers, Howard Higman, defined deference using a double negative when he said that it is the right not to know that you are not loved and cherished.

5. Garfinkel (1967) and the ethnomethodologists were particularly attuned to the way that one is best able to examine the social functioning of society when examining the breaches.

6. In the same way, a mobile phone can be seen as a personal symbol, and the style, model, and nature of the device can be interpreted (Fortunati 2003; Ling 2001).

7. These objects can include lapel pins showing a national flag, striped ties bearing the colors of a college, t-shirts and sweaters with the logo of a favorite football team, and rings indicating membership in an organization.

8. When one sees, for example, a Hopi kachina doll in a museum, one is seeing an object that is stripped of its symbolic power. Someone who has not subscribed to the culture and the situation in which that object was revered lacks access to the whole sense of the object and its ability to evoke a common sense of cohesion.

Chapter 5

1. See his analysis of ritual in the wake of the events of September 11, 2001 (Collins 2004b).

2. It is also in this study that Collins develops his ideas on co-present interaction. He noted in this study that throughout history, as communication technologies changed and developed, those who become well-known intellectuals have interacted face to face with other well-known intellectuals, often across generations. Collins's notion is that it is through intense co-present interaction that the passion and nuance of intellectual arguments can be made.

3. In his earlier work *The Sociology of Philosophies*, Collins posited a slightly different sense of Interaction ritual. The elements used there were these: (1) two or more people assembled, (2) mutual focus of attention, (3) sharing of a common mood or emotion, (4) cumulative intensification of the mood, (5) establishment of the sense that participants are members of a group with some moral obligations to each other, and (6) establishment of emotional energy (Collins 1998, pp. 22–24).

4. As examples of ritual interaction, Collins looks at sexual behavior and smoking. In both cases, there is a kind of rhythm associated with the activity. Collins has been criticized for reducing social interaction to the notion of emotional entrainment. In some cases, he suggests that it is the search for emotional energy that is the motivating factor in social interaction. This ignores the ebb and flow of demands that the small group (Fine 2005, 2006) and the exigencies of the broader society.

5. This type of situation would be an example of a failed ritual.

6. While Goffman does not exclude the potential for secondary engagements, they should be held in check lest they overtake the focus of the main engagement

(Goffman 1963a, p. 194). This theme obviously has resonance in an era in which mobile phones can threaten the focus of an engagement.

7. Indeed, this has been exploited by Joshua Meyrowitz in his analysis of broadcast communications (1985). Goffman has also been used in the analysis of interactive communication technologies by Joachim Höflich (2003) and in my own analysis of the mobile phone as a disturbance in the public sphere (2004b, 1997).

8. To be fair, Goffman also discussed failed rituals, including faux pas (1959, p. 212) and gaffes and slips made by radio announcers (1981, pp. 46, 211).

9. Of course, openings and closings are important in the management of telephonic interaction, and there is a literature on the etiquette and the functioning of telephone greetings (Sacks et al. 1974). The point with greetings and closings is that they link the current interaction to that which has taken place previously and to that which presumably will take place in the future. Thus, they can be interpreted as rituals that influence the sacred nature of the friendship or at least the relationship between individuals. The greeting takes up the thread where they had left off, and the parting is the tag that will be reactivated in future interactions.

10. It is somewhat common to use mobile phones to "transmit" concerts to friends. This, however, only describes the interaction between the concertgoer and the "interlocutor."

Chapter 6

1. Turner (1995, 95) uses 'liminal' in this context to refer to the threshold of an experience.

2. In the tradition of van Gennep's analysis of passage rites, Couldry (2003) discusses how ritual interaction necessarily requires a liminal process—that is, it needs to transport the individual across some threshold. Couldry discusses large-scale media events and the sense that, by participating, the individual is transported to another level of insight. All this must be scaled back to the small-group level and the interpersonal level when we are dealing with mobile communication.

3. There can be situations in the interpersonal world in which secondary persons take responsibility for the engineering of an inter-personal ritual. ("Harry, have you met my friend Sally? I think you two have a lot in common.") In this case, the apparatus is far more accessible than in the case of the larger scale rituals that are characteristic of Durkheim's liminal rituals.

4. According to Collins (2004a, p. 272), the participants in these events gain categorical identities.

5. Durkheim does not consider the possibility that the ritual will be unsuccessful.

6. Indeed, Collins has been criticized for seeing ritual in everything. Gary Fine (2005) notes that Collins falls into the trap of "finding emotion everywhere, watering down affect into a rather thin gruel." According to Fine, there is too

little focus on the interaction that generates solidarity and the sense of group membership.

7. Collins's point is that within the upper crust of society there is a continual round of gatherings that require a high degree of focused, scheduled, and scripted interaction, and that this results in strong group boundaries and a high degree of solidarity.

Chapter 7

1. $f(5,824) = 16.688$, sig. < 0.001.

2. See also Fortunati 2002 and Puro 2002.

3. Another issue is that the personal photograph is not necessarily totemic in the complete Durkheimian sense. There are those photos that have that status in a particular group or culture. "Moonrise over Hernandez, New Mexico" by Ansel Adams, the photographs of Babe Ruth's retirement, Nick Ut's photos of the severely burned Vietnamese children running away from napalm attacks, and similar photographs might be considered in this class at a broader social level. At the familial level there are often classic photos that capture a particularly telling moment such as a wedding or the birth of a child (Kendall 2006). These are periodically brought forward as a confirmation of some domestic event. Further, the photograph is often a referent and not a sacred object in itself. It is our discussion and interaction regarding the iconic photographs that gives us pause to reflect over various events and which gives the photos their symbolic power. These photographs touch on a similar range of feelings, the harmony of the Adams photograph, the nostalgia of Babe Ruth, or the horror of the napalm. In a minor way, the photos viewed in the observation play into the same process. Their collective viewing and the commentaries regarding them supplement the social order.

4. It has been reported that Norwegian Princess Marta-Louise sent a text message to her mother, Queen Sonja, to announce the birth of her second child. At the time she received the message, the queen was attending the funeral of Pope John Paul II.

5. The extreme example of this is when someone takes a mobile phone call in the middle of a wedding or a funeral.

6. Women waking at night in less secure areas have employed this gesture to indicate their connection to other presumably more secure realms (Ling 2004b, p. 40). The "false" use of the mobile telephone is sometimes seen as a way to avoid other engagements. Women sometimes report using the device when walking through questionable parts of town where their faked conversation is used to indicate their contact with the outside world should that be necessary. Others—significantly more often males—indicate using the same guise when trying to avoid talking to others with whom they would rather not spend the time (Baron and Ling 2007b).

7. Karl Johan Avenue, the main street through central Oslo, is used for parades and public ceremonies.

8. When walking at a saunter down the sidewalk her navigation glances were generally forward. When crossing a side street she gave lateral glances.

9. Recently I visited Tokyo and was shown the Shibuya crossing, allegedly one of the busiest pedestrian crossings in the world. With each light change, it is reported, approximately 1,500 people cross the street—or rather the streets, since there are several roads that converge there, and there is a so-called "scramble crossing" where all automobile traffic stops and only pedestrians cross the intersection. According to Rheingold (2002), 80 percent of them carry a mobile phone. In my limited experience there, despite of the crush of people, mobile phone users were able to cross the street while texting. As people cross the street, they organize themselves spontaneously into "trains" of individuals heading toward one or another side of the interaction. One person, perhaps unwittingly heads off in the appropriate direction and others simply follow him or her. The people navigate by keeping an eye on the individual immediately in front of them—usually well within a arms length. With half a mind on the person in front, they are free to concentrate on their texting. A chain of 10–15 persons can thus quickly cross the street, texting all the while.

10. I was not snoopy enough to be able to see if there were a sequence of text messages sent to a single other person or if there were several messages sent to several people.

Chapter 8

1. See also Marvin 1988.

2. For a discussion of this issue in the context of the internet, see Katz and Rice 2002.

3. The concept of a cellcert is most closely associated with the singer Clay Aiken. The point, however is that mobile telephony allows the sharing of a mass experience with others who are not present.

4. As opposed to "ahoy," suggested by Alexander Graham Bell.

5. Goffman's notion of clarifying and fixing the participants' roles seems relevant here.

6. On the acculturation of children in this special linguistic situation, see Veach 1980 and Ling and Helmerson 2000.

7. Users of the "push-to-talk" function of some mobile phones adopt this kind of interaction, since the number of persons with whom these types of conversations is quite limited and the interaction has the nature of an extended conversation.

8. This is somewhat similar to the development of the inter-office memo (Yates 1989).

9. This is a significant difference ($\chi^2(6) = 13.68$, sig. = 0.033). This material came from a survey of 1,000 randomly selected persons in Norway in 2002. Laumann et al. (1994, p. 185) report some of the same in terms of the reported number of sexual partners. Either there are differences in the gendered interpretation of

a romantic involvement (the females taking the initiative more often than the males) or the mobile phone disrupts the ability to arrive at a shared meaning of the romantic involvement.

10. Flirting and the pursuit of romance are typical adolescent uses of the mobile phone. As these individuals move into other phases of life this use of the device often recedes. In a similar way, as the adolescent finishes the emancipation process and becomes more thoroughly ensconced in adult life, some of the other more "teeny bopper" uses of the device are also shed.

11. As Ron Rice suggested in comments on a draft of this book, there may also be ritualized dimensions of romantic evaluation and rejection.

12. Obviously there are examples of relationships developing via the internet (Lea and Spears 1995), via the telegraph (Standage 1998), and also via "mail order." These are, however, extremely rare.

13. Teens also use social networking sites such as Facebook and MySpace to gather information about potential boyfriends or girlfriends and to find out about preferences and circles of friends. They also use them to make preliminary contact.

14. These examples are from a corpus of text messages gathered from a random sample of 2,003 Norwegians in May 2002 (Ling 2005c). Along with demographic, behavioral, and attitudinal questions associated with mobile telephony and SMS, we asked the respondents to read (and, where necessary, to spell out) the content of the last three messages they had sent. This resulted in a body of 882 text messages from 463 (23%) of the respondents.

15. On the use of mobile phones among Chinese for explicit but almost exclusively non-fulfilled interaction, see Law and Peng 2006.

16. Ron Rice (personal communication) reports overhearing mobile phone-based conversations in a local coffee stand where the woman called her boyfriend and reported that class was boring and further that she was very horny.

17. For an analysis of mobile communication in the Pilipino romantic life, see Solis 2007.

18. The instances discussed here come from Norway and the United States.

19. The use of argot in texting can be seen as similar to the Durkheimian totem in that it consists of symbolically imbued artifacts that bear the sense of the group between more intense interactions.

20. This is a particularly interesting (some would say jarring) juxtaposition, since it adds a post-modern element to the more traditional rural form of the word.

21. While there is a line of thought that shortened spelling is done for the sake of efficiency (Hård af Segerstad 2005b; Castells et al. 2004), there is also a line of argument that it is often done for the sake of group definition (Baron 2004; Ling 2005c; Ling and Baron 2007). The substitution of the longer word 'tjenara' when 'hei' or 'hej' would suffice indicates that in this case the need for group identification trumped efficiency.

22. When using a phone set up for the Norwegian language, the user has to cycle through the letters a, b, c, and å before coming to ä.

23. It is reported that they sometimes drop using the ä when they are in a hurry substituting it with the letter a.

24. Another way to note inclusion in Norway is to use—or eventually not use—dialect words that would underscore the writer's origin and ties to a location.

25. These citations retain the original spelling of the senders to the degree that it was possible. They come from the corpus of messages examined by Ling and Baron (2007).

26. These comments come from a group interview with teens held in 2004 in Oslo.

27. $\chi^2(36)$ = 247.22, sig. < 0.001. This and the following analysis came from a survey by the author and Telenor of 1,000 randomly selected persons in Norway in 2002.

28. $\chi^2(6)$ = 14.61, sig. = 0.023

29. These were collected in questionnaires where we asked the respondents to write down the last text messages that they had sent, not received. Thus there are there are only the outgoing text messages. This is for reasons of ethics (the person sending messages to the respondent has not agreed to participate in the data collection) and because of methodological issues (it is not possible to know the age, sex, and other socio-demographic characteristics of the individual sending the messages to the respondent).

30. The person who is thinking about telling a joke and the potential audience need to ascertain and agree upon the general context. It is bad taste to tell a joke, for example as people assemble for a funeral. If the atmospherics are in place the next task is to establish the specific context for telling a story. This can be done by hanging a comment on a previous utterance such as "that reminds me of a good one." Such a statement might also include what Goffman (1981, p. 317) calls "bracket laughs" or laughs that the teller uses to mark the beginning of a joke or a lighter phrase. This signals to others that the following utterance is an attempt to tell a joke. It sets the stage for the yarn that will follow. Interestingly this is more common for women than men (Provine 2000; Kuipers 2006). The actual process of telling a joke also has a form. In some cases (e.g., "knock, knock" jokes) it is a highly proscriptive sequence.

31. It is clear that humor can be used in large groups—e.g., via television or in theaters. The focus here, however, is on small-scale interpersonal interactions.

32. On humor in radio, see Goffman 1981.

33. In the example here, humor seemed to be an element in the interaction of a long-term dyad. Goffman describes the use of laughter to form solidarity in small ad hoc groups, such as passengers in an airport who form loose coalitions in the face of delays and other demeaning situations at the hands of the airlines. An ironic inside exchange provides the passengers insight as to their common fate—a delayed or canceled flight, for example (Goffman 1981, p. 60)

34. Of course the tables are turned, to some degree, in satire, where a less powerful person may makes jokes at the expense of a more powerful person (Holmes and Marra 2002).

35. Humor is also a way of maintaining morality in the face of unimaginable situations. Duncan (1970, p. 408) cites V. E. Frankl on the use of humor by the prisoners in German concentration camps. Humor was a "weapon of the mind in the struggle for its preservation".

36. On the centripetal forces of gossip, see Turner et al. 2003. Turner's work suggests that gossip works against cohesion. It is, however, based on experimental analysis that seeks to replicate a gossip situation in an artificial setting.

37. Humor can also function in this way, particularly when used against a group.

38. On trust in small groups, see Fine and Holyfield 1996.

39. Analysis of text messages indicates that this kind of interaction is second only to coordination messages in frequency (Ling 2005c).

Chapter 9

1. Ito (2001) reports similar behavior among teens in Tokyo.

2. Neither Putnam nor McPherson examines the role of mobile mediated interpersonal interaction on social cohesion in any detail (Boase et al. 2006).

3. This is indeed a legacy of the traditional land-line telephone, according to Thorngren (1977).

4. See also Smoreda and Thomas 2001; de Gournay and Smoreda 2003; Harper 2003.

5. In the "Ung i Norge" ("Young in Norway") study, a questionnaire was administered to 11,928 students from 47 randomly selected middle schools and 26 high schools in February 2002. The respondents were middle school and high school students. There were approximately 2,000 persons from each of the six grade levels. The age range was 13–19 years.

6. This data also show a positive correlation between the teen's sexual activity and mobile phone use (Ling 2005a; Pedersen and Samuelsen 2003) as well as narcotic misuse and restricted sleep patterns (Koivusilta et al. 2005; Punamaki et al. 2006).

7. Person correlation = 0.258, sig. = 0.013. The material presented here is based on a random sample of Norwegians (n = 1,000) interviewed in June 2006.

8. Person correlation = 0.246, sig. = 0.018.

9. Person correlation = 0.219, sig. = 0.036. These statistics relate to the use of mobile voice interaction and not texting. It is possible to speculate that texting is so common among teens that there is little variance between the more outgoing and the more reserved individuals with regards their production of text messages. Instead, it is necessary to look to voice interaction to see the differences between the more and less social teens.

10. They also report that CMC (computer-mediated communication) is used in expanding relationships with weak ties.

11. Their sample consisted of 1,050 students selected in a stratified random sampling process. The analysis looked at, among other things, mass media use, social connectedness, loneliness, and shyness.

12. The material used by Ling et al. (2003) was gathered in context of the European Union's IST project called e-living, a panel study of European (plus Israel) users. The countries included the United Kingdom, Italy, Norway, Germany, Bulgaria, and Israel. The number of users studied in the first wave (December 2001) was 10,532; the number in the second wave (December 2002) was 6,646.

13. This analysis used the Eurescom P903 multi-country survey data. The countries surveyed included Norway, Denmark, the Netherlands, Germany, the United Kingdom, France, Italy, Spain, and the Czech Republic. In each of the countries surveyed, about 1,000 individuals were sampled through regionalized random walk selection. The field work was executed in December 2000. All data are weighted for analysis so that they are representative of gender and age distribution for the population aged 15 years and older within each of the intra-national regions.

14. Such as home-based persons or those who have no fixed location of business (e.g., street hawkers).

15. See also Goodman 2005 and Horst and Miller 2005. Interestingly, Horst and Miller report that the mobile phone is used for the cultivation of contacts outside the core friendship group.

16. The material here was gathered from telephone interviews of 2,002 persons conducted in the year 2002.

17. While about half of the calls are to people within 10 kilometers, about one-fourth are to people between 10 and 50 kilometers away; the remaining one-fourth are to people farther afield. The material comes from an internet survey—with all the limitations that that implies—conducted in Norway in October 2006. The survey had 987 respondents.

18. This data comes from the Pohs survey of mobile communication in the United States conducted by Department of Communication Studies at the University of Michigan. The survey was conducted by telephone between March 3 and 10, 2005, and included 849 persons, with a response rate of 53.3%. Among other things, the survey shows that 69% of the respondents had a cell phone.

19. Slightly more than one-third of all calls are reported to be to persons within 5 miles of the caller. Another one-third go to persons who are between 5 and 25 miles away. A little more than one-fourth of calls to go to people more than 25 miles away but still within the country, and about 1% are international calls. The slightly more diffused pattern (relative to European studies) is likely to be due to the large size of US urban areas. The US system of nationwide flat rates also affects this equation.

20. $F(4,544) = 3.064$, sig. $= 0.016$.

21. The work of Miyata is a two-wave panel survey. The first wave, with 1,320 adult respondents, was conducted in November 2002 in Yamanashi prefecture in Japan. The second wave, conducted in March 2005, included 1,002 respondents who had completed the first survey. A total of 432 persons completed both surveys.

22. They also report that a preference for texting reflects social anxiety.

23. One criticism of this study is that the open internet sample makes it difficult to generalize the results. Since the sample came from an unrestricted sample of internet users, it is not possible to know to what extent the results can be generalized. The results should be read in that light.

24. The participants in this study included 132 first-year law students (64 males and 68 females) who were taking an introductory psychology class in central Japan. Data was gathered during the start of the semester for these new students and also after 12 weeks of class. The students were between 18 and 23 years of age.

25. This is the Pohs data set referred to in note 18 above.

26. Flat or quasi-flat rates for nationwide voice calls are normal in the United States. This would tip the balance in favor of voice interaction, whereas in Europe customers are often billed for actual time-based consumption.

27. This material is from a study of a random sample of 1,000 Norwegians conducted in 2005.

28. Ishii's work is based on a two-wave panel survey of randomly selected persons in Japan aged 12–69 years. The first wave included 1,878 persons surveyed in December 2001. In December 2003, the second wave was administered and resulted in completed survey instruments from 1,245 respondents.

29. This is due in part to the nature of the mobile phone and the system surrounding it. There are, for example, practical reasons that the device is not used for keeping in touch with the more far-flung and the less central persons in our social network. In many countries, the pricing of calls has traditionally been higher than with the land-line telephone—though the opposite is actually the case in the United States. Also, given the relative dearth of phone books and the rapid turnover of telephone subscriptions, it is not always possible to find the mobile phone number of an individual.

30. While the internet and IM potentially give us access to a global audience, the media make it more difficult to communicate complex or difficult messages (DiMaggio et al. 2001; Daft and Lengel 1986; Fish et al. 1992; Rice 1992; Quan-Haase and Wellman 2002). There is a diminished ability to communicate social cues and attributes of a person (e.g., gender, age, social status). Mobile communication among intimates does not rely to the same degree on these forms of interaction since the interlocutors are more familiar with each other and potentially can draw on a common reservoir insight that facilitates the interaction.

31. The flow of interactions is assisted by reliance on name registers that define the names and numbers of the group/family member (Licoppe 2004).

32. According to data gathered in Norway in 2002, 44% of teens aged 16–19 send a text message to friends via the mobile phone between midnight and 6 A.M. at least once a week. The percentage drops to 5 among 35–44-year-olds. Research

from Finland has also shown that the intensity of peer interaction via the mobile phone, particularly among teen girls can disrupt normal sleeping (Punamaki et al. 2006). The respondents to this survey were a nationally representative sample of 7,292 Finnish adolescents ages 12–18 years collected in 2001.

33. This has been confirmed by the Norwegian statistical bureau (Vaage 2005).

Chapter 10

1. This interview was conducted by my colleague Siri Nilsen.

2. Duguid (2005) describes somewhat the same issue when he discusses tacit knowledge.

3. This interview was conducted before the mobile telephone was widely adopted by teens.

4. The role of the mobile phone in the construction and maintenance of the nomos has been examined (Ling 2006).

5. Gergen calls such groups "monadic clusters," meaning clusters that cannot be reduced further.

6. Habermas (1989, n. 4) suggests something of the same with regard to computer-mediated interaction. Though computer users are more amenable to political action than simple viewers of the mass media, there is a tendency toward fractionalization into issue-related publics that have broader impact only when they crystallize around a particular topic.

7. The material cited in this paragraph comes from the Ung i Norge (Young in Norway) data described in the previous chapter. These statistics are from a study, conducted in 2002, of 11,406 teens from 73 junior high and high schools in Norway.

8. $\chi^2(5) = 32.73$, sig. < 0.001.

9. $\chi^2(5) = 23.31$, sig. < 0.001.

10. $\chi^2(5) = 17.62$, sig. $= 0.003$.

11. $\chi^2(5) = 73.41$, sig. < 0.001.

12. According to Bryant (2002), IM often includes a relatively large number of persons with whom the individual is not in daily contact. The work by Bryant, however, is based on a relatively small sample: 40 pre-teens (11–13 years old).

13. Goffman (1967, p. 95) calls this "the cult of the individual." See also Collins 2004a, pp. 370–371.

Bibliography

Aakvaag, G. 2006. Individualisering: En sosiologisk modell. *Sosiologisk tidsskrift* 14: 326–350.

Andersen, H. 2006. Melding mottatt: Ungdom og bruk av SMS i sjekkeprosessen. Institutt for sosiologi og statsvitenskap, NTNU, Trondheim, Norway.

Anderson, B. 1991. *Imagined Communities: Reflections on the Origin and Spread of Nationalism*. Verso.

Anderson, K., et al. 2005. The grandparent-grandchild relationship: Implications for models of intergenerational communication. *Human Communication Research* 31: 268–294.

Answerbag. 2006. Telephones and cell phones. http://www.answerbag.com.

Arminen, I. 2007. New reasons for mobile communication—Intensification of time-space geography in the mobile era. In *The Mobile Communications Research Annual*, volume 1, ed. R. Ling and S. Campbell. Transaction.

Aronson, S. 1977. The sociology of the telephone. *International Journal of Comparative Sociology* 12: 153–156.

Bakke, J. 1996. Competition in mobile telephony. *Telektronikk* 96: 83–88.

Banjo, O., et al. 2006. Cell Phone Usage and Social Interaction with Proximate Others: Ringing in a Theoretical Model. International Communications Association.

Barabási, A.-L. 2002. *Linked: The New Science of Networks*. Perseus.

Baron, N. 1998. Writing in the age of email: The impact of ideology versus technology. *Visible Language* 32: 35–53.

Baron, N. 2000. *Alphabet to Email: How Written English Evolved and Where It's Heading*. Routledge.

Baron, N. 2004. See you online: Gender issues in American college student use of instant messaging. *Journal of Language and Social Psychology* 23: 397–423.

Baron, N., and Ling, R. 2007. Emerging patterns of American mobile phone use: Electronically-mediated communication in transition. Conference proceedings, Association of Internet Researchers.

Bauman, Z. 2001. *The Individualized Society*. Polity.

Beck, U. 1994. The reinvention of politics: Toward a theory of reflexive modernization. In *Reflexive Modernization*, ed. U. Beck et al. Polity.

Beck, U., and Beck-Gernsheim, E. 2002. *Individualization: Institutionalized Individualism and Its Social and Political Consequences*. Sage.

Beck, U., et al. 1994. *Reflexive Modernization: Politics, Tradition and Aesthetics in the Modern Social Order*. Polity.

Bell, C. 1997. *Ritual: Perspectives and Dimensions*. Oxford University Press.

Bellah, R. 2005. Durkheim and ritual. In *The Cambridge Companion to Durkheim*, ed. J. Alexander and P. Smith. Cambridge University Press.

Berger, P., and Kellner, H. 1964. Marriage and the construction of reality. *Diogenes* 45: 1–25.

Boase, J., et al. 2006. The Strength of Internet Ties. Pew Internet and American Life Project.

Bourdieu, P. 1985. The forms of capital. In *Handbook of Theory and Research for the Sociology of Education*, ed. J. Richardson. Greenwood.

Bourdieu, P. 1995. *Distinksjonen. En sosiologisk kritikk av dømmekraften*. Pax.

Bourdieu, P. 1991. *Language and Symbolic Power*. Polity.

Bourdieu, P. 2002. *Outline of a Theory of Practice*. Cambridge University Press.

Bowles, S., and Gintis, H. 2000. Social capital and community governance. *Economic Journal* 112: F419–F436.

Brown, P., et al. 1987. *Politeness: Some Universals in Language Usage*. Cambridge University Press.

Bryant, N. 2002. The Religious Fervor of an Educational Technology Initiative: Salvation for All Classes or Opiate of the Masses? Commonwealth University, Richmond, Virginia.

Bugeja, M. 2005. *Interpersonal Divide: The Search for Community in a Technological Age*. Oxford University Press.

Burt, R. 1999. The social capital of opinion leaders. *Annals of the American Academy of Political and Social Science* 566: 37–54.

Burt, R. 2000. The network structure of social capital. In *Research in Organizational Behavior*, volume 22, ed. R. Sutton and B. Staw. JAI.

Burt, R. 2001a. Bandwidth and echo: Trust, information and gossip in social networks. In *Networks and Markets*, ed. A. Casella and J. Rauch. Russell Sage Foundation.

Burt, R. 2001b. The social capital of structural holes. In *New Directions in Economic Sociology*, ed. M. Guillien et al. Russell Sage Foundation.

Burt, R., et al. 1998. Personality correlates of structural holes. *Social Networks* 20: 63–87.

Byrne, R., and Findlay, B. 2004. Preference for SMS versus telephone calls in initiating romantic relationships. *Australian Journal of Emerging Technologies and Society* 2: 48–61.

Campbell, S. 2007. Cross cultural comparison of perceptions and uses of mobile telephony. *New Media and Society* 9, no., 2: 343–363.

Campbell, S., and Kelley, M. 2006. Mobile phone use in AA networks: An exploratory study. *Journal of Applied Communication Research* 34, no. 2: 191–208.

Campbell, S., and Kwak, N. 2007. Mobile communication and social capital in localized, globalized, and scattered networks. In proceedings of International Communications Association, San Francisco.

Campbell, S., and Russo, T. 2003. The social construction of mobile telephony: An application of the social influence model to perceptions and uses of mobile phones within personal communication networks. *Communication Monographs* 40: 317–334.

Carey, J. 1992. *Communication as Culture: Essays on Media and Society.* Routledge.

Castells, M., et al. 2004. The Mobile Communication Society: A Cross-Cultural Analysis of Available Evidence on the Social Uses of Wireless Communication Technology. Research report, Annenberg Research Network on International Communication, Los Angeles.

Cladis, M. 2005. Beyond solidarity? Durkheim and 21st century democracy in a global age. In *The Cambridge Companion to Durkheim,* ed. J. Alexander and P. Smith. Cambridge University Press.

Clark, H., and Brennan, S. 1991. Grounding in communication. In *Perspectives on Socially Shared Cognition,* ed. J. Levine and S. Teasley. American Psychological Association.

Clark, H., and Marshall, C. 1981. Definite reference and mutual knowledge. In *Elements of Discourse Understanding,* ed. A. Joshi et al. Cambridge University Press.

Clark, H., and Schaeffer, E. 1981. Contributing to discourse. *Cognitive Science* 13: 259–295.

Cohen, A., et al. 2007 *The Wonder Phone in the Land of Miracles: Mobile Telephony in Israel.* Hampton.

Coleman, J. 1988. Social capital in the creation of human capital. *American Journal of Sociology* 94: 95–120.

Collins, R. 1994. *Four Sociological Traditions.* Oxford University Press.

Collins, R. 1998. *The Sociology of Philosophies: A Global Theory of Intellectual Change.* Belknap.

Collins, R. 2004a. *Interaction Ritual Chains.* Princeton University Press.

Collins, R. 2004b. Rituals of solidarity and security in the wake of terrorist attack. *Sociological Theory* 22: 53–87.

Collins, R. 2005. The Durkheimian movement in France and in world sociology. In *The Cambridge Companion to Durkheim,* ed. J. Alexander and P. Smith. Cambridge University Press.

Costa, D., and Kahn, M. 2003. Understanding the Decline in Social Capital 1952–1998. *Kyklos* 9, no. 1: 17–46.

Couldry, N. 2003. *Media Rituals: A Critical Approach*. Routledge.

Daft, R., and Lengel, R. 1986. Organizational information requirements, media richness and structural design. *Management Science* 32: 554–571.

Danet, B. 2001. *Cyberpl@y: Communicating Online*. Berg.

Davie, R., et al. 2004. Mobile phone ownership and usage among pre-adolescents. *Telematics and Informatics* 21: 359–373.

Davis, F. 1985. Clothing and fashion as communication. In *The Psychology of Fashion*, ed. M. Solomon. Heath.

Davis, F. 1992. *Fashion, Culture, and Identity*. University of Chicago Press.

Deacon, T. 1997. *The Symbolic Species: The Co-Evolution of Language and the Brain*. Norton.

de Gournay, C. 2002. Pretense of intimacy in France. In *Perpetual Contact*, ed. J. Katz and M. Aakhus. Cambridge University Press.

de Gournay, C., and Smoreda, Z. 2003. Communication technology and sociability: Between local ties and "global ghetto"? In *Machines That Become Us*, ed. J. Katz. Transaction.

de Sola Pool, I. 1971. *The Social Impact of the Telephone*. MIT Press.

DiMaggio, P., et al. 2001. Social implications of the Internet. *Annual Review of Sociology* 27: 307–336.

Dobashi, S. 2005. The gendered use of *keitai* in domestic contexts. In *Personal, Portable, Pedestrian*, ed. M. Ito et al. MIT Press.

Dobsen, K. 2003. How Detroit police reinvented the wheel. http://www.detnews.com.

Donner, J. 2005. The Rules of Beeping: Exchanging Messages Using Missed Calls on Mobile Phones in sub-Saharan Africa. International Communications Association.

Döring, N., and Pöschl, S. 2007. Nonverbal cues in mobile phone text messages: The effects of chronemics and proxemics. In *Mobile Communications Research Annual*, volume 1, ed. R. Ling and S. Campbell. Transaction.

Douglas, M. 1966. *Purity and Danger: An Analysis of the Concepts of Pollution and Taboo*. Routledge.

Douglas, M. 2003. *Natural Symbols*. Routledge.

Douglas, M., and Isherwood, B. 1979. *The World of Goods: Toward an Anthropology of Consumption of Goods*. Routledge.

Douglas, S. 1999. *Listening In: Radio and the American Imagination*. Times Books.

Duguid, P. 2005. "The art of knowing": Social and tacit dimensions of knowledge and the limits of the community of practice. *The Information Society* 21: 109–118.

Dulaney, S., and Fiske, A. 1994. Cultural rituals and obsessive-compulsive disorder: Is there a common psychological mechanism? *Ethos* 22: 243–283.

Duncan, H. 1970. *Communication and the Social Order*. Oxford University Press.

Duncan, S. 1972. Some signals and rules for taking speaking turns in conversations. *Journal of Personality and Social Psychology* 23: 238–292.

Durkheim, E. 1951. *Suicide*. Free Press.

Durkheim, E. 1954. *The Elementary Forms of Religious Life*. Free Press.

Durkheim, E. 1995. *The Elementary Forms of Religious Life*. Free Press.

Durkheim, E. 1997. *The Division of Labor in Society*. Free Press.

Eder, D. 1988. Building cohesion through collaborative narration. *Social Psychology Quarterly* 51: 225–235.

Elliott, D., and Ageton, S. 1979. An integrated theoretical perspective on delinquent behavior. *Journal of Research in Crime and Delinquency* 16: 3–27.

Ellwood-Clayton, B. 2003. Virtual strangers: Young love and texting in the Filipino archipelago of cyberspace. In *Mobile Democracy*, ed. K. Nyiri. Passagen Verlag.

Ellwood-Clayton, B. 2005. Desire and loathing in the cyber Philippines. In *The Inside Text*, ed. R. Harper et al. Kluwer.

Farley, T. 2003. Mobile telephone history. http://www.privateline.com.

Farley, T. 2005. Privateline.com: Telephone History. http://www.privateline.com.

FCC (US Federal Communications Commission). 2005. Trends in telephone service.

Fine, G. 1979. Small groups and culture creation: The idioculture of Little League baseball teams. *American Sociological Review* 44: 733–745.

Fine, G. 1987. *With the Boys: Little League Baseball and Preadolescent Culture*. University of Chicago Press.

Fine, G. 2005. *Interaction Ritual Chains* (review). *Social Forces* 83: 1287.

Fine, G. 2006. Where the action is: Small groups and recent developments in sociological theory. *Small Group Resarch* 37: 4–19.

Fine, G., and DeSoucey, M. 2005. Joking cultures: Humor themes as social regulation in group life. *Humor* 18: 1–22.

Fine, G., and Holyfield, L. 1996. Secrecy, trust, and dangerous leisure: Generating group cohesion in voluntary organizations. *Social Psychology Quarterly* 59: 22–38.

Fischer, C. 1992. *America Calling: A Social History of the Telephone to 1940*. University of California Press.

Fischer, C. 2001. Bowling Alone: What's the Score? Proceedings of American Sociological Association.

Fish, R., et al. 1992. Video as a technology for informal communication. *Communications of the ACM* 36: 48–61.

Fishman, J. 1978. Interaction: The work women do. *Social Problems* 25: 397–406.

Flugel, J. C. 1950. *The Psychology of Clothes*. Hogarth.

Fortunati, L. 2000. The mobile phone: New social categories and relations. In *Sosiale konsekvenser av mobiletelefoni*, ed. R. Ling and K. Thrane. Telenor FoU.

Fortunati, L. 2002. Italy: Stereotypes: True or false. In *Perpetual Contact*, ed. J. Katz and M. Aakhus. Cambridge University Press.

Fortunati, L. 2005. Is body-to-body communication still the prototype? *The Information Society* 21: 53–61.

Fortunati, L. 2003. Mobile phone and the presentation of self. Presented at Front Stage–Back Stage conference, Grimstad, Norway.

Fortunati, L., and Manganelli, A. 2002. El teléfono móvil de los jóvenes. *Revista de Estudios de Juventud* 57: 59–78.

Fortunati, L., et al. 2003. *Mediating the Human Body: Technology, Communication and Fashion*. Erlbaum.

Fox, J. 1996. How does civil society thicken? The political construction of social capital in rural Mexico. *World Development* 24: 1089–1103.

Franklin, M., et al. 2000. Effectiveness of exposure and ritual prevention for obsessive—Compulsive disorder randomized compared with nonrandomized samples. *Journal of Consulting and Clinical Psychology* 68: 594–602.

Freeman, A. 1993. *God in Us: A Case for Christian Humanism*. SCM Press.

Friedkin, N.E. 2004. Social cohesion. *Annual Review of Sociology* 30: 409–425

Freud, S. 1963. Obsessive acts and religious practice. In *Character and Culture*, ed. P. Reiff. Collier Books.

Garfinkel, H. 1967. *Studies in Ethnomethodology*. Basic Books.

Garton, L., et al. 1997. Studying Online Social Net-works. http://jcmc.indiana.edu.

Gergen, K. 2003. Self and community and the new floating worlds. In *Mobile Democracy*, ed. K. Nyri. Passagen Verlag.

Gergen, K. 2008. Mobile communication and the transformation of democratic process. In *Handbook of Mobile Communication Studies*, ed. J. Katz. MIT Press.

Giddens, A. 1986. *The Constitution of Society: Outline of the Theory of Structuration*. University of California Press.

Giddens, A. 1994. Risk, trust, reflexivity. In *Reflexive Modernization*, ed. U. Beck et al. Polity.

Gluckman, M. 1963. Gossip and scandal. *Current Anthropology* 4: 307–316.

Goffman, E. 1959. *The Presentation of Self in Everyday Life*. Doubleday Anchor Books.

Goffman, E. 1961. *Asylums: Essays on the Social Situation of Mental Patients and Other Inmates*. Doubleday Anchor Books.

Goffman, E. 1963a. *Behavior in Public Places: Notes on the Social Organization of Gatherings*. Free Press.

Goffman, E. 1963b. *Stigma: Notes on the Management of Spoiled Identity.* Simon and Schuster.

Goffman, E. 1967. *Interaction Ritual: Essays on Face-to-Face Behavior.* Pantheon.

Goffman, E. 1971. *Relations in Public: Microstudies of the Public Order.* Harper.

Goffman, E. 1981. *Forms of Talk.* University of Pennsylvania Press.

Goffman, E., and Loflund, L. 1989. On fieldwork: Transcription of a talk given at the 1974 Pacific Sociological Meetings. *Journal of Contemporary Ethnography* 18: 123–132.

Goffman, E., and Verhoeven, J. 1993. An interview with Erving Goffman 1980. *Research on Language and Social Interaction* 26: 317–348.

Goggin, G. 2006. Notes on the history of the mobile phone in Australia. *Southern Review* 38: 4–22.

Goodman, J. 2005. Linking Mobile Phone Ownership and Use to Social Capital in Rural South Africa and Tanzania. http://www.vodafone.com.

Granovetter, M. 1973. The strength of weak ties. *American Journal of Sociology* 78: 1360–1380.

Grenier, P., and Wright, K. 2001. Social capital in Britain: Update and critique of Peter Hall's analysis. In proceedings of ARNOVA conference, Miami.

Grinter, R., and Eldridge, M. 2001. Y do tngrs luv 2 txt msg? In *Proceedings of the Seventh European Conference on Computer Supported Cooperative Work,* ed. W. Prinz et al. Kluwer.

GSM World. 2007. Messaging. http://www.gsmworld.com.

Habermas, J. 1989. *The Theory of Communicative Action.* Beacon.

Habuchi, I. 2005. Accelerating reflexivity. In *Personal, Portable, Pedestrian,* ed. M. Ito et al. MIT Press.

Haddon, L. 2001. Domestication and mobile telephony. In *Machines That Become Us,* ed. J. Katz. Transaction.

Haddon, L. 2004. *Information and Communication Technologies in Everyday Life.* Berg.

Hall, E. 1973. *The Silent Language.* Doubleday Anchor Books.

Hall, P. 1999. Social capital in Britan. *British Journal of Politics* 29: 417–461.

Hård af Segerstad, Y. 2005a. Language in SMS—A socio-linguistic view. In *The Inside Text,* ed. R. Harper et al. Springer.

Hård af Segerstaad, Y. 2005b. Language use in Swedish mobile text messaging. In *Mobile Communications,* ed. R. Ling and P. Pedersen. Springer.

Hardin, G. 1968. The tragedy of the commons. *Science* 162: 1243–1248.

Harper, R. 2003. Are mobiles good or bad for society? In *Mobile Democracy,* ed. K. Nyiri. Passagen Verlag.

Harvey, A. 2006. The liminal magic circle: Boundaries, frames, and participation in pervasive mobile games. http://www.wi-not.ca/.

Hatch, M., and Ehrlich, S. 1993. Spontaneous humour and an indicator of paradox and ambiguity in organizations. *Organization Studies* 14: 505–527.

Haythornthwaite, C. 2001. Exploring multiplexity: Social network learning structures in a computer-supported distance learning class. *The Information Society* 17: 211–226.

Hjorth, L. 2006. Fast-forwarding present: The rise of personalisation and customisation in mobile technologies in Japan. *Southern Review* 38: 23–42.

Hjorthol, R. 2000. Same city–different options: An analysis of the work trips of married couples in the metropolitan area of Oslo. *Journal of Transportation Geography* 8: 213–220.

Höflich, J. 2003. *Mensch, Computer und Kommunikation*. Peter Lang.

Holmes, J. 2006. Sharing a laugh: Pragmatic aspects of humor and gender in the workplace. *Journal of Pragmatics* 38: 26–50

Holmes, J., and Marra, M. 2002. Over the edge? Subversive humor between colleagues and friends. *Humor* 15: 65–85.

Horst, H., and Miller, D. 2005. From kinship to link-up: Cell phones and social networking in Jamaica. *Current Anthropology* 46: 755–778.

Humphreys, L. 2005. Social topography in a wireless era: The negotiation of public and private space. *Journal of Technical Writing and Communication* 35: 367–384.

Igarashi, T., et al. 2005. A longitudinal study of social network development via mobile phone text messages focusing on gender differences. *Journal of Social and Personal Relationships* 22: 691–713.

Imray, L., and Middleton, A. 1983. Public and private: Marking the boundaries. In *In The Public and the Private*, ed. E. Gamarnikow et al. Heinemann.

Ishii, K. 2006. Implications of mobility: The uses of personal communication media in everyday life. *Journal of Communications* 56: 346–365.

Ito, M. 2001. Mobile phones, Japanese youth and the re-placement of social contact. In *Mobile Communications*, ed. R. Ling and P. Pedersen. Springer.

Ito, M. 2003. Camera phones changing the definition of picture-worthy. http://www.japanesemediareview.com.

Ito, M. 2004. Personal, portable, pedestrian: Lessons from Japanese mobile phone use. In *Mobile Communication and Social Change*, ed. S. Kim. SK Telecom.

Ito, M., et al. 2005a. *Personal, Portable, Pedestrian*, ed. M. Ito et al. MIT Press.

Ito, M. 2005b. Mobile phones, Japanese youth and the replacement of social contact. In *Mobile Communications*, ed. R. Ling and P. Pedersen. Springer.

Ito, M., and Okabe, D. 2005. Technosocial situations: Emergent structuring of mobile e-mail use. In *Personal, Portable, Pedestrian*, ed. M. Ito et al. MIT Press.

Ito, M., and Okabe, D. 2006. Intimate connections: Contextualizing Japanese youth and mobile messaging. In *Information Technology at Home*, ed. R. Kraut. Oxford University Press.

ITU (International Telecommunication Union). 2005. Mobile Cellular, Subscribers per 100 People. http://www.itu.int.

Jaworski, A., and Coupland, J. 2005. Othering in gossip: "You go out you have a laugh and you can pull yeah okay but like. . . ." *Language in Society* 34: 667–694.

Jessop, G. 2006. A brief history of mobile telephony: The story of phones and cars. *Southern Review* 38: 43–60.

John, P., et al. 2003. Social capital and causal role of socialization. Presented at ESRC Democracy and Participation conference, University of Essex.

Johnsen, T. 2000. Ring meg! En studie av ungdom og mobiltelefoni. Department of Ethnology, University of Oslo.

Johnstone, A., et al. 1995. There was a long pause: Influencing turn taking behavior in human-human and human-computer spoken dialogues. *International Journal of Human Computer Studies* 41: 383–411.

Jones, D. 1980. Gossip: Notes on women's oral culture. *Women's Studies International Quarterly* 3: 193–198.

Jones, R. A. 1986. *Emile Durkheim: An Introduction to Four Major Works.* Sage.

Katz, J., and Aakhus, M. 2002a. *Perpetual Contact: Mobile Communication, Private Talk, Public Performance.* Cambridge University Press.

Katz, J., and Aakhus, M. 2002b. Conclusion: Making meaning of mobiles—a theory of *apparatgeist.* In *Perpetual Contact,* ed. J. Katz and M. Aakhus. Cambridge University Press.

Katz, J., and Aspden, P. 1997. A nation of strangers? *Communications of the ACM* 40: 81–86.

Katz, J., and Rice, R. 2002. *Social Consequences of Internet Use.* MIT Press.

Katz, J., et al. 2004. Personal mediated communication and the concept of community in theory and practice. *Communication Yearbook* 28: 315–371.

Kavanaugh, A., and Patterson, S. 2001. The impact of community computer networks on social capital and community involvement. *American Behavioral Scientist* 45: 496–509.

Kavanaugh, A., and Reese, D. 2005. Weak ties in networked communities. *The Information Society* 21: 119–131.

Kawachi, B., et al. 1999. Social capital and self rated health: A contextual analysis. *American Journal of Public Health* 89: 1187–1193.

Kazmer, M., and Haythornthwaite, C. 2001. Juggling multiple social worlds. *American Behavioral Scientist* 45: 510–529.

Kearney, M. 2005. Birds on the wire: Troping teenage girlhood through telephony in mid-twentieth-century US media culture. *Cultural Studies* 19: 568–601.

Kendall, L. 2006. Something old ... Remembering Korean wedding hall photographs from the 1980s. *Visual Anthropology* 19: 1–19.

Kendon, A. 1967. Some functions of gaze-direction in social interaction. *Acta Psychologica* 26: 26–63.

Kim, H. 2006. The configurations of social relationships in communication channels: F2F, email, messenger, mobile phone, and SMS. Presented at International Communications Association pre-conference on mobile communication.

Kim, S. 2002. Korea: Personal meanings. In *Perpetual Contact*, ed. J. Katz and M. Aakhus. Cambridge University Press.

Koivusilta, L., et al. 2005. Intensity of mobile phone use and health compromising behaviours—How is information and communication technology connected to health-related lifestyle in adolescence? *Journal of Adolescence* 28: 35–47.

Kraut, R., et al. 1998. Internet paradox: A social technology that reduces social involvement and psychological well being? *American Psychologist* 53: 1017–1031.

Kuipers, G. 2006. *Good Humor, Bad Taste: A Sociology of the Joke*. Mouton de Gruyter.

Lampert, M., and Ervin-Tripp, S. 2002. Risky laughter: Teasing and self-directed joking among male and female friends. *Journal of Pragmatics* 38: 51–72.

Lash, S. 1994. Reflexivity and its doubles: Structure, aesthetics and community. In *Reflexive Modernization*, ed. U. Beck et al. Polity.

Lash, S. 2002. Individualization in a non-linear mode. In *Individualization*, ed. U. Bech and E. Beck-Gernsheim. Sage.

Latham, R., and Sassen, S. 2004. *Digital Formations: IT and New Architectures in the Global Realm*. Princeton University Press.

Laumann, E., et al. 1994. *The Social Organization of Sexuality*. University of Chicago Press.

Laurier, E. 2001. Why people say where they are during mobile phone calls. *Environment and Planning* 19: 485–504.

Law, P.-L., and Peng, Y. 2006. The use of mobile phones among migrant workers in southern China. In *New Technologies in Global Societies*, ed. P.-L. Law et al. World Scientific.

Lea, M., and Spears, R. 1995. Love at first byte? Building personal relationships over computer networks. In *Under-Studied Relationships*, ed. J. Wood and S. Duck. Sage.

Lerum, K. 2004. Sexuality, power, and camaraderie in service work. *Gender and Society* 18: 756–776.

Licoppe, C. 2004. Connected presence: The emergence of a new repertoire for managing social relationships in a changing communications technoscape. *Environment and Planning D: Society and Space* 22: 135–156.

Licoppe, C., and Heurtin, P. 2001. Managing one's availability to telephone communication through mobile phones: A French case study of the development dynamics of the use of mobile phones. *Personal and Ubiquitous Computing* 5: 131–140.

Lindmark, S. 2002. Evolution of Techno-Economic Systems: An Investigation of the History of Mobile Communications. Doctoral dissertation, Chalmers University of Technology, Gothenberg, Sweden.

Ling, R. 1997. "One can talk about common manners!" The use of mobile telephones in inappropriate situations. In *Themes in Mobile Telephony*, ed. L. Haddon. Telia.

Ling, R. 1998. "She Calls, [but] It's for Both of Us You Know": The Use of Traditional Fixed and Mobile Telephony for Social Networking among Norwegian Parents. Telenor R&D, Kjeller.

Ling, R. 1999. "We release them little by little": Maturation and gender identity as seen in the use of mobile telephone. In proceedings of International Symposium on Technology and Society, Rutgers University.

Ling, R. 2000. Direct and mediated interaction in the maintenance of social relationships. In *Home Informatics and Telematics*, ed. A. Sloane and F. van Rijn. Kluwer.

Ling, R. 2001. "It is 'in.' It doesn't matter if you need it or not, just that you have it.": Fashion and the domestication of the mobile telephone among teens in Norway. In *Il corpo umano tra tecnologie, comunicazione e moda*, ed. L. Fortunati. Triennale di Milano.

Ling, R. 2004a. "Goffman er gud": Thoughts on Goffman's usefulness in the analysis of mobile telephony. Wireless Communication Workshop, Department of Communication, University of Michigan, Ann Arbor.

Ling, R. 2004b. *The Mobile Connection: The Cell Phone's Impact on Society*. Morgan Kaufmann.

Ling, R. 2004c. Where is mobile communication causing social change? In *Mobile Communication and Social Change*, ed. S. Kim. Korean Association of Broadcasting Studies.

Ling, R. 2005a. Mobile communications vis-à-vis teen emancipation, peer group integration and deviance. In *The Inside Text*, ed. R. Harper et al. Kluwer.

Ling, R. 2005b. The socio-linguistics of SMS: An analysis of SMS use by a random sample of Norwegians. In *Mobile Communications*, ed. R. Ling and P. Pedersen. Springer.

Ling, R. 2005c. The socio-linguistics of SMS: An analysis of SMS use by a random sample of Norwegians. In *Mobile Communications*, ed. R. Ling and P. Pedersen. Springer.

Ling, R. 2006. Flexible coordination in the Nomos: Stress, emotional maintenance and coordination via the mobile telephone in intact families. In *Cultural Dialectics and the Cell Phone*, ed. A. Kavoori and N. Arceneaux. Peter Lang.

Ling, R. 2007a. Informal social capital and ICTs. In *Information and Communications Technologies in Society*, ed. B. Anderson et al. Routledge.

Ling, R. 2007b. Exclusion of elderly persons in the case of text messaging. In *Personlige medier*, ed. L. Prøitz and T. Rasmussen. University of Oslo.

Ling, R., ed. 2004d. Report of Literature and Data Review, Including Conceptual Framework and Implications for IST. European Union.

Ling, R., and Baron, N. 2007. The mechanics of text messaging and instant messaging among American college students. *Journal of Sociolinguistics,* in press.

Ling, R., and Helmersen, P. 2000. "It must be necessary, it has to cover a need": The adoption of mobile telephony among pre-adolescents and adolescents. In *The Social Consequences of Mobile Telephony*. Telia.

Ling, R., et al. 2002. E-living deliverable 6. Family, gender and youth: Wave one analysis. IST.

Ling, R., et al. 2003. Mobile communication and social capital in Europe. In *Mobile Democracy*, ed. K. Nyri. Passagen Verlag.

Ling, R., et al. 2005. Nascent communication genres within SMS and MMS. In *The Inside Text*, ed. R. Harper et al. Kluwer.

Logica CMG. 2006. Message Plus: Fact sheet.

Lohan, E. 1996. The domestic phone space: The telephone space in the home and the domestic space in telephony a participant observation case study of the telephone in three Dublin households. Report prepared for COST 248 Project.

Love, S. 2001. Space invaders: Do mobile phone conversations invade peoples' personal space? In *Human Factors in Telecommunications*, ed. K. Nordby. HFT.

Love, S. 2005. *Understanding Mobile Human-Computer Interaction.* Butterworth Heinemann.

Lynne, A. 2000. Nyansens makt–en studie av ungdom, identitet og klær. Statens institutt for forbruksforskning, Lysaker.

Maclean, A., and Yocom, J. 2000. Interview with Randall Collins. http://www .ssc.wisc.edu.

Manning, P. 1996. Information technology in the police context: The "sailor" phone. *Information Systems Research* 7: 52–62.

Martin, M. 1991. *Hello, Central?: Gender, Technology and Culture in the Formation of Telephone Systems.* McGill–Queens University Press.

Marvin, C. 1988. *When Old Technologies Were New: Thinking about Electric Communication in the Late Nineteenth Century.* Oxford University Press.

Marx, K. 1995. *The Poverty of Philosophy.* Prometheus.

McIntosh, W., and Harwood, P. 2002. The Internet and America's changing sense of community. *The Good Society* 11: 25–28.

McPherson, M., et al. 2006. Social isolation in America: Changes in core discussion networks over two decades. *American Sociological Review* 71: 353–375.

Merton, R. 1968. *Social Theory and Social Structure.* Free Press.

Meyrowitz, J. 1985. *No Sense of Place: The Impact of Electronic Media on Social Behavior.* Oxford University Press.

Miyata, K. 2006. Longitudinal Effects of Mobile Internet Use on Social Network in Japan. In proceedings of International Communications Association conference, Dresden.

Monk, A., et al. 2004. Why are mobile phones annoying? *Behavior and Information Technology* 23: 33–41.

Morgan, S., and Sorensen, A. 1999. Parental networks, social closure, and mathematical learning: A test of Coleman's social capital explanation of school effects. *American Sociological Review* 64: 661–681.

Murtagh, G. 2002. Seeing the "rules": Preliminary observations of action, inter-action and mobile phone use. In *Wireless World*, ed. B. Brown et al. Springer.

National Statistical Coordination Board (Philippines). 2007. ICT Statistics: Show me the data! http://www.nscb.gov.

Nie, N. 2001. Sociability, interpersonal relations, and the Internet: Reconciling conflicting findings. *American Behavioral Scientist* 45: 420–435.

Nie, N., et al. 2002. Internet use, interpersonal relations and sociability: A time diary study. In *The Internet in Everyday Life*, ed. B. Wellman and C. Haythornthwaite. Blackwell.

Nordal, K. 2000. Takt og tone med mobiltelefon: Et kvalitativt studie om folks brug og opfattelrer af mobiltelefoner. Institutt for sosiologi og samfunnsgeografi, Universitetet i Oslo.

Okabe, D., and Ito, M. 2005. *Keitai* in public transportation. In *Personal, Portable, Pedestrian*, ed. M. Ito et al. MIT Press.

Palen, L., et al. 2001. Discovery and integration of mobile communications in everyday life. *Personal and Ubiquitous Computing* 5: 109–122.

Park, W. 2005. Mobile telephone addiction. In *Mobile Communications*, ed. R. Ling and P. Pedersen. Springer.

Pedersen, W., and Samuelsen, S. 2003. Nye mønstre av seksualatferd blant ung-dom. *Tidsskrift for Den norske lægeforeningen* 21: 3006–3009.

Portes, A. 1998. Social capital: Its origins and applications in modern sociology. *Annual Review of Sociology* 24: 1–24.

Preece, J. 2004. Etiquitte online: From nice to necessary. *Communications of the ACM* 47: 56–61.

Prøitz, L. 2004. The mobile fiction: Intimate discourses in text message communi-cation amongst young Norwegian people. In *Mobile Communication and Social Change*, ed. S. Kim. SK Telecom.

Prøitz, L. 2005. Intimacy fiction: Intimate discourses in mobile telephone commu-nication amongst Norwegian youth. In *A Sense of Place*, ed. K. Nyiri. Passagen Verlag.

Prøitz, L. 2006. Cute Boys or Game Boys? The Embodiment of Femininity and Masculinity in Young Norwegian's Text Message Love-Projects. http://journal .fibreculture.org.

Provine, R. 2000. *Laughter: A Scientific Investigation*. Penguin.

PT (Post- og Teletilsyn, Oslo). 2004. Det Norske Telemarked: 1. halvår 2004.

Punamaki, R.-L., et al. 2006. Use of information and communication technol-ogy (ICT) and perceived health in adolescence: The role of sleeping habits and waking-time tiredness. *Journal of Adolescence* 27: 1–17.

Puro, J.-P. 2002. Finland: a mobile culture. In *Perpetual Contact*, ed. J. Katz and M. Aakhus. Cambridge University Press.

Putnam, R. 1995. Bowling alone: America's declining social capital. *Journal of democracy* 6: 65–78.

Putnam, R. 2000. *Bowling Alone: The Collapse and Revival of American Community*. Touchstone.

Quan-Haase, A., and Wellman, B. 2002. Capitalizing on the Net: Social contact, civic engagement and sense of community. In *The Internet in Everyday Life*, ed. B. Wellman and C. Haythornthwaite. Blackwell.

Rafael, V. 2003. The cell phone and the crowd: Messianic politics in the contemporary Philippines. *Public Culture* 15: 399–425.

Rakow, L. 1988. Women and the telephone: The gendering of a communications technology. In *Technology and Women's Voices*, ed. C. Kramarae. Routledge.

Rakow, L. 1992. *Gender on the Line*. University of Illinois Press.

Rawls, A. 1996. Durkheim's epistemology: The neglected argument. *American Journal of Sociology* 102: 430–482.

Rawls, A. 2003. Conflict as a foundation for consensus: Contradictions of industrial capitalism in book III of Durkheim's *Division of Labor*. *Critical Sociology* 29: 295–335.

Reid, D., and Reid, F. 2004. Insights into the Social and Psychological Effects of SMS Text Messaging. http://www.160characters.org.

Rheingold, H. 2002. *Smart Mobs*. Perseus.

Rice, R. 1987. Computer-mediated communication and organizational innovation. *Journal of Communications* 37: 65–94.

Rice, R. 1992. Task analyzability, use of new media, and effectiveness: A multi-site exploration of media richness. *Organization Science* 3: 475–500.

Rice, R., and Katz, J. 2003. Mobile discourtesy: National survey results on episodes of convergent public and private spheres. In *Mobile democracy: Essays on society, self and politics*, ed. K. Nyíri. Passagen Verlag.

Riesman, D., N. Glazer, and R. Denney. 1950. *The Lonely Crowd: A Study in the Changing American Character*. Yale University Press.

Rivère, C. 2002. La pratique du mini-message, une double strataie d'exterioisation et de retrait de l'intimite dans les interactions quotidiennes. *Reseaux* 20: 112–113.

Rivère, C., and Licoppe, C. 2005. From voice to text: Continuity and change in the use of mobile phones in France and Japan. In *The Inside Text*, ed. R. Harper et al. Springer.

Sacks, H., et al. 1974. The simplest systematics for the organization of turn-taking for conversations. *Language* 50: 696–735.

Salen, K., and Zimmerman, E. 2004. *Rules of Play: Game Design Fundamentals*. MIT Press.

Sattel, J. 1976. The inexpressive male: Tragedy or sexual politics. *Social Problems* 23: 469–477.

Scheff, T. 1990. *Microsociology: Discourse, Emotion and Social Structure*. University of Chicago Press.

Schegloff, E. 1998. Body torque. *Social Research* 65: 535–596.

Schegloff, E., and Sacks, H. 1973. Opening up closings. *Semiotica* 8: 289–327.

Schegloff, E., et al. 1977. The preference for self correction in the organization of repair in conversation. *Language* 53: 361–382.

Schwartz, G., and Merten, D. 1967. The language of adolescents: An anthropological approach to the youth culture. *American Journal of Sociology* 72: 453–468.

Searle, J., and Freeman, A. 1995. The construction of social reality: Anthony Freeman in conversation with John Searle. *Journal of Consciousness Studies* 2: 180–189.

Selberg, T. 1993. Television and the ritualization of everyday life. *Journal of Popular Culture* 26: 3–11.

Selberg, T. 1995. Fjernsynsvirkelighet og hverdagsvirkelighet. Om bruk av fjernsyn i den norske hverdagen. In *Nostalgi og sensasjoner*, ed. T. Selberg. NIF Publications.

Sennett, R. 1998. *The Corrosion of Character: The Personal Consequences of Work in the New Capitalism*. Norton.

Shiu, E., and Lenhart, A. 2004. How Americans use instant messaging. http://www.pewInternet.org.

Silverstone, R., et al. 1992. Information and communication technologies and moral economy of the household. In *Consuming Technologies*, ed. R. Silverstone and E. Hirsch. Routledge.

Simmel, G. 1949. The sociology of sociability. *American Journal of Sociology* 3: 254–261.

Simmel, G. 1971. *Georg Simmel: On Individuality and Social Forms*. University of Chicago Press.

Slater, P. 1963. On social regression. *American Sociological Review* 28: 339–64.

Smardon, R. 2005. Where the action is: The microsociological turn in educational research. *Educational Researcher* 34, no. 1: 20–25.

Smoreda, Z., and Thomas, F. 2001. Social networks and residential ICT adoption and use. In proceedings of EURESCOM Summit 2001. EURESCOM.

Solis, R. 2007. Texting love: An exploration of text messaging as a medium for romance in the Philippines. *M/C Journal* 10, no. 1. http://journal.media-culture.org.

Spencer, B., and Gillen, F. 1899. *The Northern Tribes of Central Australia*. Macmillan.

Standage, T. 1998. *The Victorian Internet*. Weidenfeld and Nicolson.

Sykes, A. 1966. Joking relationships in an industrial setting. *American Anthropologist* New Series 68: 188–193.

Tannen, D. 1991. *You Just Don't Understand: Men and Women in Conversation*. Virago.

Taylor, A. 2005. Phone-talk and local forms of subversion. In *Mobile Communications*, ed. R. Ling and P. Pedersen. Springer.

Thomas, W. 1931. *The Child in America*. Knopf.

Thorngren, B. 1977. Silent actors: Communication networks for development. In *The Social Impact of the Telephone*, ed. I. de Sola Pool. MIT Press.

Tortora, P., and Eubank, K. 1989. *A Survey of Historic Costume*. Fairchild Publications.

Travers, J., and Milgram, S. 1969. An experimental study of the small world problem. *Sociometry* 32: 425–443.

Treichler, P., and Kramarae, C. 1983. Women's talk in the ivory tower. *Communication Quarterly* 31: 118–132.

Trosby, F. 2004. SMS, the strange duckling of GSM. *Telektronikk* 3: 187–194.

Turkle, S. 1995. *Life on the Screen: Identity in the Age of the Internet*. Simon and Schuster.

Turner, M., et al. 2003. Relational ruin or social glue? The joint effect of relationship type and gossip valence on liking, trust, and expertise. *Communication Monographs* 70: 129–141.

Turner, V. 1995. *The Ritual Process: Structure and Anti-Structure*. Aldine.

Tönnies, F. 1965. Gemeinschaft and Gesellschaft. In *Theories of Society*, ed. T. Parsons et al. Free Press.

University of Michigan. 2006. On the Move: The Role of Cellular Communication in American Life. Department of Communication Studies, Ann Arbor.

Vaage, O. 2005. Mediabruks undersøkelse. Statistics Norway.

van Gennep, A. 2004. *The Rites of Passage*. Routledge.

Veach, S. 1980. Children's Telephone Conversations. Ph.D. dissertation, Stanford University.

Warner, W. 1957. *A Black Civilization: A Social Study of an Australian Tribe*. Harper.

Warner, W., et al. 1963. *Yankee City*. Yale University Press.

Watkins, E. 2005. Instant fame: Message boards, mobile phones, and Clay Aiken. Presented at "Internet Research 6.0: Internet Generations" conference, Chicago.

Weber, M. 2002. *The Protestant Ethic and the Spirit of Capitalism*. Routledge.

Wei, R., and Lo, V. 2006. Staying connected while on the move: Cell phone use and social connectedness. *New Media and Society* 8: 53–72.

Wellman, B., et al. 2001. Does the Internet increase, decrease or supplement social capital? *American Behavioral Scientist* 45: 436–455.

Williams, A., and Thurlow, C. 2005. *Talking Adolescence: Perspectives on Communication in the Teenage Years*. Peter Lang.

Wolak, J., Mitchell, K., and Finkelhor, D. 2003. Escaping or connecting? Characteristics of youth who form close online relationships. *Journal of Adolescence* 26: 105–119.

Woolcock, M. 2001. The place of social capital in understanding social and economic outcomes. Presented at Symposium on the Contribution of Human and Social Capital to Sustained Economic Growth and Well Being, Quebec.

Woolgar, S. 2003. Mobile back to front: Uncertainty and danger in the theory-technology relation. In *Mobile Communications*, ed. R. Ling and P. Pedersen. Springer.

Wurtzel, A., and Turner, C. 1977. Latent functions of the telephone: What missing the extension means. In *The Social Impact of the Telephone*, ed. I. de Sola Pool. MIT Press.

Yates, J. 1989. *Control through Communication: The Rise of System in American Management*. Johns Hopkins University Press.

Index